STUDENT UNIT GUIDE

CCEA AS Physics Unit 2

Waves, Photons and Medical Physics

Caroline Greer

PHILIP ALLAN

Philip Allan, an imprint of Hodder Education, an Hachette UK company, Market Place, Deddington, Oxfordshire OX15 0SE

Orders
Bookpoint Ltd, 130 Milton Park, Abingdon, Oxfordshire OX14 4SB
tel: 01235 827827
fax: 01235 400401
e-mail: education@bookpoint.co.uk
Lines are open 9.00 a.m.–5.00 p.m., Monday to Saturday, with a 24-hour message answering service. You can also order through the Philip Allan Updates website: www.philipallan.co.uk

ISBN 978-1-4441-8288-0

First printed 2013
Impression number 5 4 3 2 1
Year 2018 2017 2016 2015 2014 2013

Cover photo: Photodisc/Cadmium

Typeset by Integra Software Services Pvt. Ltd., Pondicherry, India

Printed in Italy

Hachette UK's policy is to use papers that are natural, renewable and recyclable products and made from wood grown in sustainable forests. The logging and manufacturing processes are expected to conform to the environmental regulations of the country of origin.

Contents

Content Guidance

Questions & Answers

Getting the most from this book

Questions & Answers

Exam-style questions

Examiner comments on the questions
Tips on what you need to do to gain full marks, indicated by the icon ⓔ.

Sample student answers
Practise the questions, then look at the student answers that follow each set of questions.

Examiner commentary on sample student answers
Find out how many marks each answer would be awarded in the exam and then read the examiner comments (preceded by the icon ⓔ) following each student answer.

Questions & Answers

Self-assessment test I

Question I

This is a graph of a progressive wave travelling at a speed of 12 cm s⁻¹.

(a) State the amplitude of the wave. (1 mark)
(b) Determine the wavelength of the wave. (2 marks)
(c) What is the phase difference between points X and Y on the wave? (1 mark)
(d) (i) Define the periodic time of a wave. (1 mark)
    (ii) Calculate the periodic time of the wave. (3 marks)

Total: 8 marks

ⓔ If a question includes a graph you must first check the label on each axis. This is a displacement–distance graph. Do not confuse this with a displacement–time graph. The wavelength can be obtained directly from a displacement–distance graph by determining the distance between successive crests. Periodic time cannot be obtained directly from a displacement–distance graph.

Answer to Question I
(a) 1.8 cm ✓

ⓔ The amplitude is the maximum displacement from the mid-point of the oscillation, so in this graph it is the displacement from the horizontal axis to the top of a crest. It is not from a crest to a trough as this is twice the amplitude. Read the scale carefully.

About this book

This guide is one of a series covering the CCEA specification for AS and A2 physics. It offers advice for the effective revision of **Unit AS 2: Waves, Photons and Medical Physics**. Its aim is to help you understand the physics and give you guidance on the core aspects of the subject. The guide has two sections:

- **Content Guidance** — this section is not intended to be a detailed textbook. It offers guidance on the main areas of the content of Unit AS 2 and includes worked examples. These examples illustrate the types of question that you are likely to come across in the examination.
- **Questions and Answers** — this comprises two self-assessment tests. Answers are provided and there is an indication of the specific points for which marks are awarded.

Physics is not an easy subject, but by committing time and effort to understanding the key elements of the discipline you can maximise your performance in the examination. The development of an understanding of physics can only evolve with experience and practice. This guide will facilitate your progress by focusing on the essential components and providing examples for you to attempt before learning from the answers.

The specification

The CCEA specification is a detailed statement of the physics that is required for the unit assessments, and describes the format of the assessments. It can be obtained from the CCEA website at **www.rewardinglearning.org.uk**.

Your teacher may have introduced you to concepts outside the specification to further develop your physics. The purpose of this book is to help you prepare for the Unit AS 2 examination.

Content Guidance

Waves

- A wave is a disturbance that propagates through a medium.
- Waves carry energy.
- There are many ways to classify waves.

Progressive and standing waves

Progressive waves transfer energy from one place to another. They are sometime called travelling waves as they appear to move. Examples of progressive waves ar electromagnetic waves travelling from the Sun to Earth or a sound wave travellin from a loudspeaker to an ear.

Standing waves do not involve the transmission of energy, as the wave energy i stored in the system. They are sometimes called stationary waves as they do no appear to move. Examples of standing waves are the vibrations in a violin string c air vibrating inside a flute.

Mechanical and electromagnetic waves

Mechanical waves are produced by a disturbance in a material and are transmitte by the oscillating particles of the material. Examples of mechanical waves are wate waves or waves on a slinky spring

Electromagnetic waves consist of oscillating electric and magnetic fields. Example include any member of the electromagnetic spectrum.

Transverse and longitudinal waves

A **transverse wave** is one in which the vibrations of the particles are at righ angles to the direction in which the wave travels. Examples are water waves c electromagnetic waves.

A **longitudinal wave** is one in which the direction of the vibrations is parallel to th direction in which the wave travels. Examples include sound waves. The back-and forth oscillations of the air particles form regions of high and low pressure calle **compressions** and **rarefactions** respectively.

Longitudinal and transverse waves can be demonstrated using a slinky sprin (Figure 1).

Longitudinal and transverse waves can be represented on graphs. For a longitudina wave (Figure 2a) it is necessary to look at how the molecules of the medium are disturbec For a transverse wave (Figure 2b) the graph looks very similar to the actual wave.

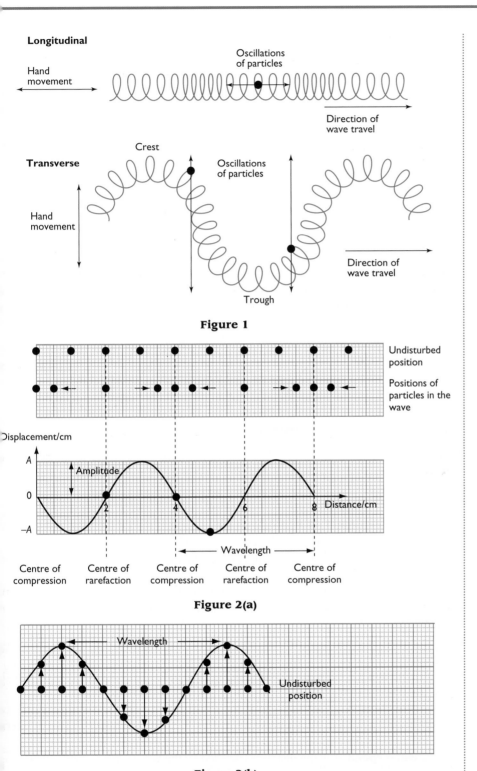

Longitudinal

Hand movement

Oscillations of particles

Direction of wave travel

Transverse

Crest

Oscillations of particles

Hand movement

Direction of wave travel

Trough

Figure 1

Undisturbed position

Positions of particles in the wave

Displacement/cm

A

Amplitude

0

−A

Distance/cm

Wavelength

Centre of compression

Centre of rarefaction

Centre of compression

Centre of rarefaction

Centre of compression

Figure 2(a)

Wavelength

Undisturbed position

Figure 2(b)

Examiner tip

It is easy to confuse these graphs. Always check which graph you are using by looking carefully at the labels on the axes.

Knowledge check 2

How can frequency be determined from a displacement–time graph? Can frequency be determined from a displacement–distance graph?

Examiner tip

Care should be taken to distinguish between displacement and amplitude. Amplitude is the *maximum* displacement.

Examiner tip

Periodic time and frequency are related. A frequency of 10 Hz means 10 oscillations per second or one oscillation is completed in $\frac{1}{10}$ second. So the periodic time of this oscillation is 0.1 s.

$$\text{frequency} = \frac{1}{\text{period}}$$

Examiner tip

Equations must have the same units on both sides. This is a good check that the equation you are using is correct.

Knowledge check 3

Calculate the frequency of a wave with a periodic time of 1 ms.

Wave properties

The oscillating particles of a wave can be represented on two types of graph. Th displacement–time graph for a single particle within the wave shows how th displacement of this particle from its equilibrium position varies with time (Figure 3,

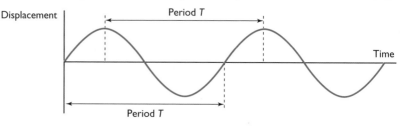

Figure 3

A displacement–distance graph shows the position of all the particles in a section o the wave at a single instant (Figure 4).

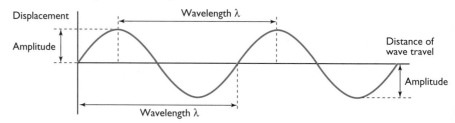

Figure 4

Definitions

- The **displacement**, x, of a particle on a wave is its distance from the mid-point o the oscillation. The unit of displacement is the metre (m).
- The **amplitude**, A, of the wave is the maximum displacement of a particle from the mid-point of the oscillation. The unit is the metre (m).
- A **crest** is a point on the wave of maximum displacement in the positive direction
- A **trough** is a point on the wave of maximum displacement in the negative direction.
- The **periodic time**, T, or **period** is the time taken for one complete oscillation o the wave. The unit is the second (s).
- The **frequency**, f, is the number of complete waves that pass a point in one second or the number of oscillations per second. The unit is the hertz (Hz).
- The **wavelength**, λ, of the wave is the distance between consecutive points o corresponding phase. It can be measured as the distance between two adjacen crests or between two adjacent troughs. The unit is the metre (m).
- The **wave speed**, v, is the distance travelled by the wave each second. The unit is metres per second (m s^{-1}).
- The **wave equation** relates the speed of a wave to its frequency and wavelength A wave travels one wavelength in the time taken for one oscillation.

As speed is the distance moved per unit time,

$$v = \frac{\lambda}{T} = \frac{1}{T} \times \lambda = f \times \lambda$$

The wave equation is:

velocity = frequency × wavelength
$$v = f\lambda$$

where v is the speed of the wave in metres per second (m s^{-1}), f is the frequency of the wave in hertz (Hz) and λ is the wavelength in metres (m).

Worked example

A girl generates a wave on a rope by moving her hand up and down, as shown in Figure 5. She generates two waves every second.

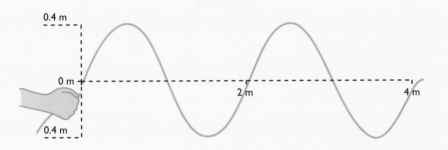

Figure 5

(a) What types of wave is she generating?

(b) What is the amplitude of the waves?

(c) What is the frequency of the waves?

(d) What is the speed of the waves?

Answer

(a) Transverse waves, as the vibrations are at right angles to the direction of the wave.
Mechanical waves, as the energy is transmitted by the oscillating particles of the rope.
Progressive waves, as the energy is being transferred from one place to another.

(b) 0.4 m (amplitude is the maximum displacement from the mid-point of the wave)

(c) Frequency = 2 waves per second = 2 Hz (frequency is the number of waves per second)

(d) λ = 2 m (wavelength is the distance between two crests)
$$v = f \times \lambda = 2 \times 2 = 4 \, \text{m s}^{-1}$$

Examiner tip
Frequency and wavelength are inversely proportional. Lower-frequency waves have longer wavelengths. Higher-frequency waves have shorter wavelengths.

Examiner tip
Always check that the units are correct and consistent. Speed and wavelength must use the same units of length. Questions involving the wave equation often involve units with prefixes such as MHz, kHz, μm and nm. Learn all the prefixes for multiples and sub-multiples of units and practise putting them into your calculator.

Examiner tip
Always state units correctly using a negative index, not a slash. For example, the unit of velocity is m s^{-1} *not* m/s.

Examiner tip
When answering questions involving calculations follow this four-point plan:
1 Write down the **formula** you are going to use.
2 **Substitute** the quantities into the formula, making sure you have consistent units for the quantities.
3 **Calculate** the answer.
4 Include the correct **unit** with your answer.

Knowledge check 4

A bat emits a sound pulse of wavelength 5.0 mm and frequency 68 kHz. Calculate the speed of the emitted sound.

Phase and phase difference

Phase describes the particular point in the cycle of a wave. It is used to compare the motion of vibrating particles in a wave or waves. Consider the motion of a single particle with the displacement–time graph in Figure 6.

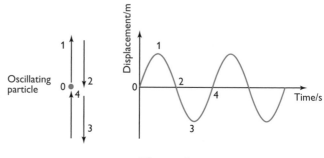

Figure 6

The points labelled 0, 1, 2, 3 and 4 represent the position of the particle at different stages of one complete cycle of the oscillation. One full cycle or one oscillation of a wave (from point 0 to point 4) is considered to be 360° or 2π radians.

- The timing of the oscillation begins at point 0 when the particle is passing up through the mid-point of the oscillation.
- At point 1, the particle is at the positive maximum displacement and has completed one quarter of the cycle of the oscillation. This stage is out of phase with point 0 as it is different by one quarter of the cycle of the oscillation. It is 90° or $\frac{\pi}{2}$ radians out of phase with point 0.
- At point 2, the particle is moving down through the midpoint of the oscillation. This is the same displacement as at point 0, but not moving in the same direction. It has completed half of the oscillation. It is 180° or π radians out of phase with point 0. This is known as antiphase.
- At point 3, the particle is at the negative maximum displacement and has completed three-quarters of the oscillation. It is 270° or $\frac{3\pi}{2}$ out of phase with point 0.
- At point 4, the particle has completed one oscillation and is at the same position and moving in the same direction as at point 0. It is in phase with point 0.

Consider the motion of the particles in a wave in the displacement–distance graph in Figure 7.

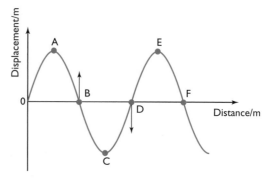

Figure 7

Now we are comparing the motion of different particles at a single instant along a section of a wave. Particle A is in phase with particle E, as it is the same displacement from the mid-point of its oscillation and moving in the same direction. Particle B is in antiphase with particle D as it is π radians out of phase, but it is in phase with particle F.

We can also use phase to compare two waves with the same frequency (Figure 8). One wave is leading the other wave, or one wave is lagging behind the other.

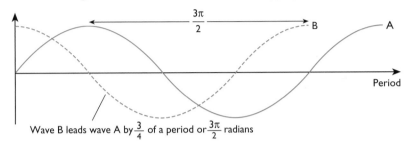

Wave B leads wave A by $\frac{3}{4}$ of a period or $\frac{3\pi}{2}$ radians

Figure 8

The phase difference is calculated between any two waves of the same frequency by finding the fraction of the complete 2π radians that represents the difference in phase. The two waves are out of step by a time t. The phase difference is equal to $\frac{t}{T} \times 2\pi$ or $\frac{t}{T} \times 360°$.

Worked example

For each of the graphs in Figure 9 state the phase difference in words, in terms of wavelength, in degrees and in radians.

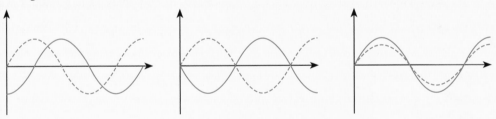

Figure 9

Answer

(a) phase difference = 90° or $\frac{\lambda}{4}$ or $\frac{\pi}{2}$ — out of phase

(b) phase difference = 180° or $\frac{\lambda}{2}$ or π — completely out of phase

(c) phase difference = 360° or λ or 2π — in phase

Polarisation

The condition for a wave to be plane-polarised is for the oscillations to be in just one plane. This phenomenon only occurs with transverse waves, for example light waves.

In normal, unpolarised light waves, such as light from a filament lamp, electric field oscillations occur in all planes perpendicular to the direction of travel of the wave. If this unpolarised light is passed through a polarising filter (called Polaroid), the

oscillations in all planes but one will be absorbed. The light emerging from the filter has oscillations in one plane only and is said to be plane-polarised.

A second sheet of Polaroid can be used to confirm that the light is plane-polarised. The second sheet is called the analyser and it is held in line with the polariser. The analyser is rotated through 360° and the light intensity will alternate between maximum and minimum (extinction) light intensities every 90°. The intensity depends on whether the transmission axis of the analyser is parallel or perpendicular to the transmission axis of the polariser (Figure 10).

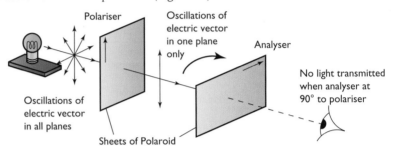

Figure 10

Note: Longitudinal waves cannot be plane-polarised as the oscillations are parallel to the direction of wave travel.

Checking that a wave is polarised

An analyser is used to check if waves are polarised. An analyser produces a polarised wave itself. If the original beam is unpolarised, as the analyser is rotated it will continually polarise the waves in successive planes as it is rotated. There will always be one plane that remains. Hence the intensity remains constant (Figure 11).

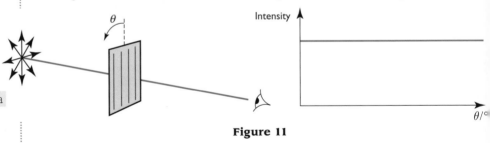

Figure 11

If the original beam is polarised, as the analyser is rotated it will vary from a maximum, when the analyser is aligned with the plane of polarisation, to a minimum when the analyser is at right angles to the plane of polarisation (Figure 12).

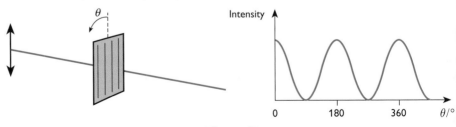

Figure 12

Examiner tip

Learn the experimental procedure to plane-polarise waves and the procedure to check that a wave is plane-polarised.

Examiner tip

When describing polarisation you must describe the oscillations being confined to one plane. The word 'plane' is a key word in the definition.

Examiner tip

Ultrasound cannot be polarised as it is a longitudinal wave.

CCEA AS Physics

Worked example

A microwave generator produces plane-polarised microwaves. An aerial connected to an amplifier and ammeter can be used to analyse the polarisation of the waves. When the aerial is parallel to the plane of polarisation, the maximum signal is received. As the aerial is rotated the signal intensity reduces, reaching a minimum when the aerial has rotated through 90°.

(a) What is meant by plane-polarised microwaves?

(b) Draw a labelled diagram of the apparatus you would use to demonstrate that the microwaves are plane-polarised.

(c) What does this experiment demonstrate about the nature of microwaves?

(d) Sketch a graph to show how the intensity of the signal changes as the aerial is rotated.

(e) Explain why sound waves cannot be plane-polarised.

Answer

(a) Plane-polarised microwaves are ones in which the oscillations are confined to one plane only, perpendicular to the direction of the wave travel.

(b)

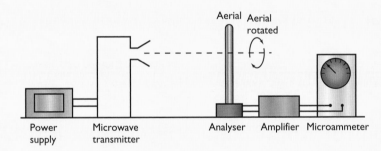

(c) As microwaves are plane-polarised the experiment shows they are transverse waves.

(d)

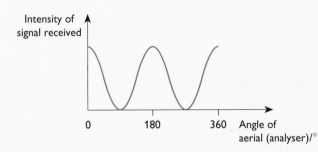

(e) Sound is a longitudinal wave and longitudinal waves cannot be plane-polarised as the oscillations are parallel to the direction of wave travel.

Knowledge check 6

Describe the variation in light intensity as an analyser is rotated through 360° while viewing polarised light.

Examiner tip

Always draw labelled diagrams and graphs using a ruler.

Electromagnetic waves

Electromagnetic waves cover a large range of wavelengths called the electromagnetic spectrum (Table 1). This continuous spectrum is divided into regions of waves depending on how the waves are produced. These broad regions overlap; there is no clear boundary between regions.

All electromagnetic waves:
- are transverse waves
- consist of oscillating electric and magnetic fields
- can travel through a vacuum
- travel at a speed of $3.00 \times 10^8 \, \mathrm{m \, s^{-1}}$ in a vacuum

Table 1 Electromagnetic waves and their applications

Region of the electromagnetic spectrum	Typical wavelength range/m	Applications
Gamma rays	10^{-16} to 10^{-12}	Sterilising medical equipment Medical imaging Killing cancerous cells
X-rays	10^{-12} to 10^{-9}	Airport security scanners Medical imaging — detecting broken bones Studying crystal structures
Ultraviolet	10^{-9} to 10^{-7}	Vitamin D production in the body Sun beds Causes fluorescence in security marking and detecting forged bank notes
Visible light	4×10^{-7} (violet) to 7×10^{-7} (red)	Photosynthesis Photography Vision
Infrared	10^{-6} to 10^{-4}	Cooking, grilling Remote controls Thermal imaging devices
Microwaves	10^{-4} to 10^{-1}	Cooking Mobile phones Satellite communications
Radio	10^{-1} to 10^{5}	Radio and television communications Radio astronomy

Examiner tip

Learn the order of the electromagnetic spectrum and know which is the high-frequency end and which is the long-wavelength end.

Knowledge check 7

State **three** properties of radio waves and explain why they are not considered as being dangerous.

Examiner tip

Use the correct units in the formula. Convert frequency to hertz. Convert wavelength to metres. A common mistake is to use kHz or cm.

Worked example

The speed of electromagnetic waves in air is given by $c = 3 \times 10^8 \, \mathrm{m \, s^{-1}}$.

(a) Calculate the wavelength of radio waves of frequency 250 kHz.

(b) Calculate the frequency of microwaves of wavelength 5 cm.

Answer

(a) $v = f\lambda$ so $\lambda = \dfrac{v}{f} = \dfrac{c}{f} = \dfrac{3 \times 10^8}{2.5 \times 10^5} = 1200 \, \mathrm{m}$

(b) $v = f\lambda$ so $f = \dfrac{v}{\lambda} = \dfrac{c}{\lambda} = \dfrac{3 \times 10^8}{0.05} = 6 \times 10^9 \, \mathrm{Hz}$

- Waves can be classified as transverse or longitudinal. In transverse waves, the vibrations are perpendicular to the direction of wave travel. In longitudinal waves, the vibrations are parallel to the direction of wave travel.

- There are certain characteristics that describe a wave. The displacement of a particle is the distance from the mid-point of the oscillation and the amplitude is the maximum displacement. The periodic time is the time taken for one complete oscillation of the wave. The frequency is the number of complete waves that pass a point in one second. The wavelength is the distance between consecutive points of corresponding phase.

- The speed of a wave is the product of its frequency and wavelength, $v = f\lambda$. This is called the wave equation. Periodic time is related to frequency by $T = 1/f$.

- A displacement–time graph shows how the displacement of a particle varies with time. Periodic time can be determined from this graph. A displacement–distance graph shows the position of all particles in a section of the wave at a single instant. Wavelength can be determined from this graph.

- Phase describes the particular point in the cycle of a wave. Phase difference can be used to compare different points in a wave or points in different waves. Points are in phase if separated by a whole number of complete cycles of the wave. Points are completely out of phase if separated by an odd number of half cycles of the wave.

- A wave is plane-polarised when the oscillations occur in one plane only. Transverse waves can be polarised. Light and all the electromagnetic waves can be polarised. Longitudinal waves, for example sound, cannot be polarised. A polariser is used to polarise waves, such as Polaroid film for light. An analyser can be used to test for polarisation, for example another piece of Polaroid film for polarised light. As the analyser is rotated the light intensity varies and is extinguished twice in a 360° rotation.

- The electromagnetic spectrum is a map of all types of light. It includes visible light and other types that the eye cannot detect. The spectrum can be listed in order of increasing wavelength or frequency. The most energetic waves have the highest frequency and shortest wavelength. The electromagnetic spectrum in order of increasing frequency is: radio waves, microwaves, infrared radiation, visible light (red–violet), ultraviolet radiation, X-rays, gamma rays. For electromagnetic waves the speed of the electromagnetic waves in a vacuum is $c = 3 \times 10^8\,\mathrm{m\,s^{-1}}$. So the wave equation becomes $c = f\lambda$ for electromagnetic waves.

Refraction

Waves travel at different speeds in different media. For example, sound travels at approximately $340\,\mathrm{m\,s^{-1}}$ in air, $1500\,\mathrm{m\,s^{-1}}$ in water and $5000\,\mathrm{m\,s^{-1}}$ in steel.

Refraction occurs when a wave changes direction on travelling between different media due to a change in speed. All waves can be refracted. For example, as water waves move from deep to shallow water there may be a change of direction (Figure 13).

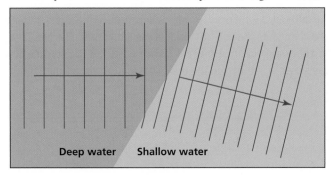

Deep water   **Shallow water**

Figure 13

If a wave slows down as it passes from one medium to the next, it will bend towards the normal (Figure 14a). If the wave speeds up as it passes from one medium to the next, it will bend away from the normal (Figure 14b). If the wave meets the boundary between the two media along the normal, at right angles to the surface, then the wave will change speed but there will be no bending (Figure 14c).

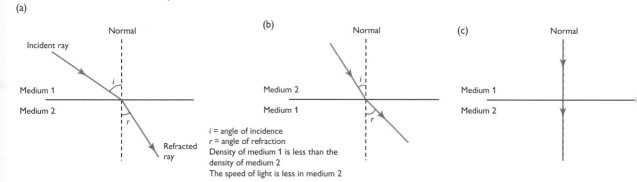

i = angle of incidence
r = angle of refraction
Density of medium 1 is less than the density of medium 2
The speed of light is less in medium 2

Figure 14

Examiner tip

In all diagrams showing refraction draw a normal, as all angles are measured from the normal.

Refraction can be shown by a ray of light passing through a parallel-sided glass block (Figure 15).

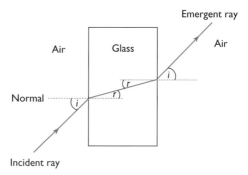

Figure 15

Note that the emergent ray is parallel to the incident ray.

Explanation of refraction

Light travels at almost $3 \times 10^8 \, \text{m s}^{-1}$ in air. When it enters a denser material, it slows down. If the whole ray hits the new material at the same time, the whole ray slows down together and no bending occurs. However, if the ray strikes the surface at an angle, then some of the ray slows down before the rest. This leads to bending (Figure 16).

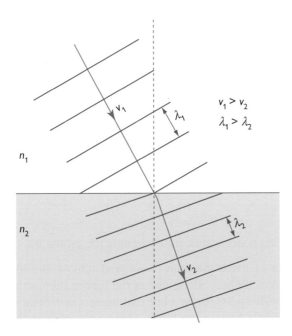

Figure 16

Refractive index

Refraction occurs at the boundary between two media because of the change of speed of the wave on entering the second medium. The ratio of the speed of the wave in the first medium v_1 to the speed of the wave v_2 in the second medium is called the **refractive index** for the boundary between those two media. It is denoted by $_1n_2$:

$$_1n_2 = \frac{\text{speed in medium 1}}{\text{speed in medium 2}} = \frac{v_1}{v_2}$$

Snell's law

Snell's law states that the ratio of the sine of the angle of incidence to the sine of the angle of refraction is the same for all rays travelling across a given boundary.

This means that the ratio $\frac{\sin i}{\sin r}$ is a constant for any two given media. This ratio for the boundary between any two particular media is also equal to the refractive index.

Snell's law is useful as it enables us to predict where a ray of light will go when it enters a medium.

Note that $_1n_2$ is the relative refractive index for light travelling between medium 1 and medium 2. If medium 1 is a vacuum then the refractive index is called the absolute refractive index. The absolute refractive indices of a pair of media determine their relative refractive index:

$$_1n_2 = \frac{n_2}{n_1} = \frac{\sin i}{\sin r}$$

If the path of the light were reversed and it travelled from medium 2 to medium 1 then the refractive index would be written as $_2n_1$:

$$_1n_2 = \frac{1}{_2n_1}$$

Experiment to verify Snell's law

Apparatus: drawing board, paper, transparent block, protractor, ruler, ray box

Draw the outline of the transparent block and remove the block. Place an X at a point a third of the way along one of the longer sides. Construct a normal (perpendicular) at this point. Draw a line that intersects the block at X at an angle of 20° to the normal. Replace the block. Shine a ray of light along the line and mark the ray that emerges from the other side of the block. Remove the block and complete the lines to show the incident, refracted and emergent rays. The path of the ray through the block is complete. Measure the angle of refraction in the block and record the result. Repeat the procedure for angles of incidence of approximately 30°, 40°, 50° and 60° and record the results. Find the sine of the angles of incidence and the sine of the angles of refraction.

The results are recorded in a table such as this:

Angle of incidence $i/°$	Angle of refraction $r/°$	$\sin i$	$\sin r$
20.0	13.5	0.34	0.23
30.0	20.5	0.50	0.35
40.0	27.0	0.64	0.45
49.0	33.5	0.76	0.55
58.0	38.0	0.85	0.62

Snell's law gives:

$$n = \frac{\sin i}{\sin r}$$

Rearranging:

$$\sin i = n \sin r$$

Comparing with the straight line equation:

$$y = mx + c$$

If a graph is plotted of $\sin i$ on the y-axis against $\sin r$ on the x-axis, a straight-line graph through the origin will verify Snell's law (Figure 17). The gradient of the straight line is equal to the refractive index n.

Examiner tip

If a question requires you to describe an experiment, draw a labelled diagram with a ruler, include a procedure and how you will use the results to form a conclusion.

Examiner tip

Results tables must have units. The units must follow the quantity at the top of each column and a slash must be used to separate the quantity from the corresponding unit (not brackets).

CCEA AS Physics

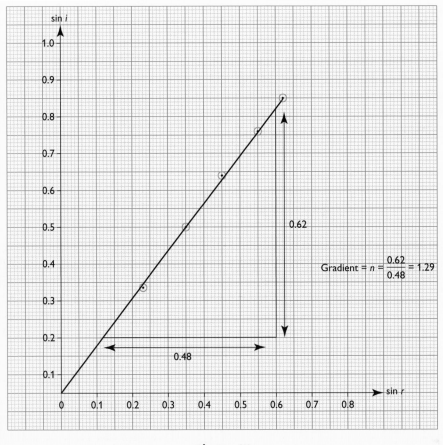

Figure 17

Examiner tip

To confirm which graph to plot, rearrange the formula and do a direct mapping to the equation of a straight line, $y = mx + c$ (m is the gradient and c is the y-intercept).

Knowledge check 9

Explain how you could demonstrate that quantities you have plotted on a graph have a proportional relationship.

Examiner tip

For refraction and reflection, angles are always measured from a normal. A normal is necessary as not all surfaces are flat. Some surfaces are irregular and some are curved, making an accurate value of an angle measured from the surface difficult to achieve.

Worked example

Light is incident on a layer of oil on the surface of a tank of water at an angle of 57°. If the refractive index of oil is 1.26 and that of water is 1.3 calculate:

(a) the angle of refraction in the oil

(b) the angle of refraction in the water

(c) the angle of refraction in the water if the oil is removed

(d) Comment on the answers to parts (b) and (c).

Answers

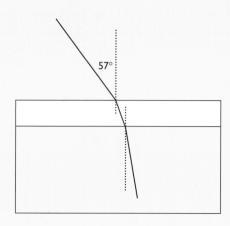

(a) angle of incidence = i = 57°

$$\frac{\sin i}{\sin r} = 1.26$$

$$\sin r = \frac{\sin i}{1.26} = \frac{\sin 57}{1.26} = 0.67$$

$$r = 41.7°$$

(b) $_{\text{oil}}n_{\text{water}} = \dfrac{n_{\text{water}}}{n_{\text{oil}}} = \dfrac{\sin i}{\sin r}$

$$\frac{1.3}{1.26} = \frac{\sin i}{\sin r} = \frac{\sin 41.7}{\sin r}$$

$$\sin r = 0.64$$
$$r = 40.15°$$

(c) i = 57°

$$\frac{\sin i}{\sin r} = \frac{\sin 57}{\sin r} = n = 1.3$$

$$\sin r = 0.64$$
$$r = 40.15°$$

(d) The answers are the same because the parallel-sided layer of oil will only laterally displace the light and make no difference to the angle at which the ray enters the water.

Total internal reflection

When light travels from a material of high refractive index to a material of lower refractive index, such as from glass to air, it bends away from the normal (Figure 18).

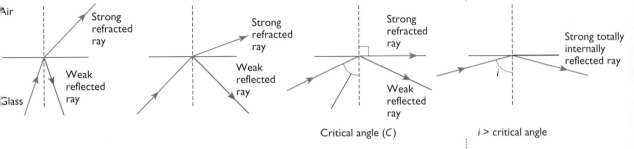

Figure 18

For small angles of incidence, most of the light will be refracted out of the glass and there will be weak reflection at the boundary. As the angle of incidence is increased, the angle of refraction will increase. When the angle of refraction is 90°, the angle of incidence is called the **critical angle**, C. When the angle of incidence is greater than the critical angle, all the light is reflected at the boundary and total internal reflection occurs.

Conditions for total internal reflection:
- Light is travelling from a material of high refractive index to a material of lower refractive index.
- The angle of incidence in the material of higher refractive index is greater than the critical angle for that material.

Examiner tip
Total internal reflection only takes place when going from a less dense to a more dense material.

Critical angle

The critical angle is the angle of incidence that produces an angle of refraction of 90°. The value of the critical angle depends on the refractive index of the material.

Applying Snell's law to light travelling from glass to air, we get:

$$_{glass}n_{air} = \frac{\sin i}{\sin r} = \frac{\sin C}{\sin 90} = \frac{\sin C}{1} = \sin C$$

but

$$_{glass}n_{air} = \frac{1}{_{air}n_{glass}}$$

so

$$\sin C = \frac{1}{_{air}n_{glass}}$$

or

$$_{air}n_{glass} = \frac{1}{\sin C}$$

Content Guidance

Experiment to measure the refractive index of glass using total internal reflection

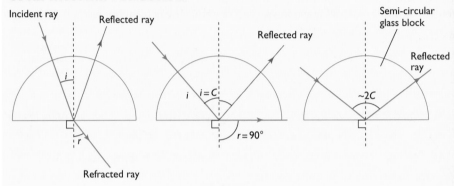

Figure 19

A semi-circular glass block is used to ensure no refraction occurs at the air–glass boundary, as the light is directed along a radius and hence along the normal (Figure 19). The angle of incidence *i* at the glass–air boundary is increased until the critical angle is reached and the angle of refraction is 90°. The light travels along the straight edge of the block. The angle of incidence can be measured at this point and is the critical angle. Alternatively, for ease of measurement, the angle of incidence could be increased by an extremely small amount so that total internal reflection takes place. The weak reflected ray becomes very bright. The angle between the incident ray and this bright reflected ray is approximately 2C. Once the critical angle is measured, the refractive index for glass can be found using:

$$_{air}n_{glass} = \frac{1}{\sin C}$$

Applications of total internal reflection

Optical fibres use total internal reflection to send light along a long, thin strand of glass, which has a core surrounded by a cladding of lower refractive index. Light is confined to the fibre as long as it is incident on the core–cladding boundary at an angle greater than the critical angle (Figure 20).

Many optical instruments, such as binoculars and telescopes, use 45° prisms to reflect light rather than mirrors, which can cause multiple reflections and blurring (Figure 21).

Examiner tip

The critical angle is not the angle of reflection when total internal reflection is occurring. The critical angle is the angle of incidence when the angle of refraction is 90°.

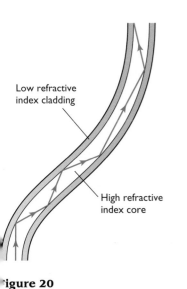

Low refractive index cladding

High refractive index core

Figure 20

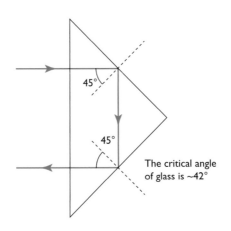

45°

45°

The critical angle of glass is ~42°

Figure 21

Knowledge check 10

Calculate the speed of light travelling through an optical fibre made of glass with a refractive index of 1.5.

Knowledge check 11

State **two** reasons why telecommunications use optical fibres rather than copper cables to transmit signals.

Summary

- At the boundary between two different media an incident wave can be reflected, transmitted (possibly refracted) or absorbed. The angle of incidence is the angle between the incident ray and the normal. The angle of reflection is the angle between the reflected ray and the normal. The angle of refraction is the angle between the transmitted ray and the normal.

- The law of reflection always demands that the angle of incidence equals the angle of reflection.

- Refraction is the change of direction of a wave caused by a change in speed as it enters a medium with different density. If the wave meets the boundary between the two materials at an angle of incidence greater than 0° (not along the normal), then the wave bends towards the normal when entering a more dense material and away from the normal when entering a less dense material. If the wave meets the boundary along the normal, there is no refraction.

- Refraction is described by Snell's law, which states that the ratio of the sine of the angle of incidence to the sine of the angle of refraction is always the same at the boundary between two particular media.

- The ratio is called the refractive index and is also equal to the ratio of the speeds in the two media:

$$_1n_2 = \frac{\sin i}{\sin r} = \frac{v_1}{v_2}$$

- Total internal reflection occurs when light travels from a more dense to a less dense material and the angle of incidence is greater than the critical angle. The critical angle is the angle of incidence for which the angle of refraction is 90°. Refractive index and critical angle are related by $n = \frac{1}{\sin C}$.

- The principle of total internal reflection is used in communications with fibre optic cables, medical endoscopes and reflectors using 45° prisms.

Lenses

A lens is a piece of transparent material with at least one curved surface, which works by refracting light. There are two types of lens (Figure 22).

Converging lens (convex lens)   **Diverging lens (concave lens)**

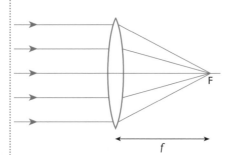 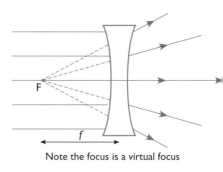

Note the focus is a virtual focus

Figure 22

Definitions

Examiner tip

Remember that a real image is one that the rays of light actually pass through and can be formed on a screen. A virtual image is one that the rays of light only appear to pass through and cannot be formed on a screen.

Knowledge check 12

Define the principal focus of a converging lens.

- A **convex lens** is thicker in the middle than at the edges and refracts parallel rays of light to converge to a real principal focus.
- A **concave lens** is thicker at the edges than in the middle and refracts parallel rays of light to diverge from a virtual principal focus.
- The **principal axis** of a lens is the line joining the centres of curvature of its two surfaces.
- The **principal focus**, F, of a lens is the point on the principal axis towards which all rays parallel to the principal axis converge in the case of a convex lens or from which they appear to diverge in the case of a concave lens, after refraction.
- Light can fall on either surface of a lens and so a lens has two principal foci, one on each side.
- The distance between the centre of the lens and the principal focus is the **focal length**, f, of the lens.

Ray diagrams for a converging lens

To find the image of an object, two of the following three rays must be constructed (Figure 23):
- A ray parallel to the principal axis is refracted through the principal focus.
- A ray through the centre of the lens is undeviated.
- A ray through the principal focus is refracted parallel to the principal axis.

(a)

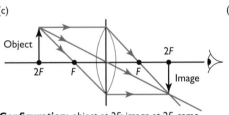

Configuration: object at infinity: point image at *F*

Applications: burning a hole with a magnifying glass

(b)

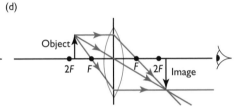

Configuration: object outside 2*F*; image between *F* and 2*F*, diminished, inverted, real

Applications: lens of a camera, human eyeball lens

(c)

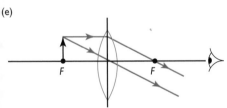

Configuration: object at 2*F*; image at 2*F*, same size as object, inverted, real

Applications: inverting lens of a field telescope

(d)

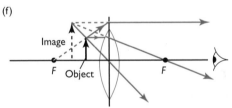

Configuration: object between *F* and 2*F*; image outside 2*F*, magnified, inverted, real

Applications: slide projector and objective lens in a compound microscope

Knowledge check 13

Describe in full the image formed by a magnifying glass.

(e)

Configuration: object at *F*; image at infinity

Applications: lenses used in lighthouses and searchlights

(f)

Configuration: object inside *F*; image on the same side of the lens as the object, magnified, upright, virtual

Applications: magnifying glass, binoculars

Figure 23

Ray diagram for a diverging lens

Figure 24 shows how to draw a ray diagram for a diverging lens.

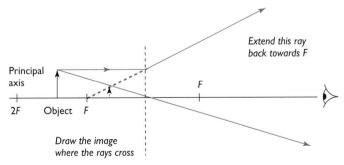

Figure 24

Lens formula

The position of an image can be calculated using the lens formula.

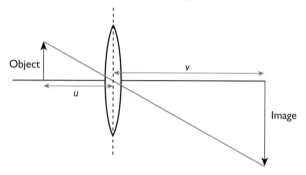

Figure 25

To use this formula the following are defined (Figure 25):
- The distance of the object from the centre of the lens is called **object distance**, u.
- The distance of the image from the centre of the lens is called the **image distance**, v.
- The distance of the principal focus from the centre of the lens is the **focal length**, f.
- The focal length of a converging lens is positive and the focal length of a diverging lens is negative.
- Distances from the centre of the lens to real images are positive and to virtual images are negative.

The lens formula is:
$$\frac{1}{f} = \frac{1}{u} + \frac{1}{v}$$

Worked example

Use the lens formula to find the position of an image formed when an object is placed 20 cm from a converging lens of focal length 15 cm.

Answer

Object distance = u = 20 cm

Image distance = v = ?

Focal length = f = 15 cm

$$\frac{1}{f} = \frac{1}{u} + \frac{1}{v}$$

$$\frac{1}{v} = \frac{1}{f} - \frac{1}{u} = \frac{1}{15} - \frac{1}{20} = \frac{1}{60}$$

$$v = 60\,\text{cm}$$

Examiner tip

Make sure when you write the quantities u and v that you form the letters carefully so that the two are not confused.

Examiner tip

A common error is to forget to calculate the reciprocal of $\frac{1}{v}$ to find v.

Examiner tip

The sign of the values is very important in the lens formula. If the image distance is positive, the image is real and is formed on the opposite side of the lens from the object.

Knowledge check 14

A virtual image is formed 8 cm from a convex lens of an object 4 cm from the lens. Calculate the focal length of the lens.

Experiment to measure the focal length of a converging lens

An *approximate* value for the focal length of a converging lens can be obtained using the light from a distant object (e.g. a window). The converging lens will form the image of the distant object at the principal focus. The lens is held so that a sharp image of the distant object is formed on a screen. The distance between the lens and the screen is measured and this is the focal length of the lens.

A more *accurate* method uses an illuminated object aligned with the lens and a screen on an optical bench or with a metre rule. The object is often a piece of wire mesh or a metal washer (Figure 26).

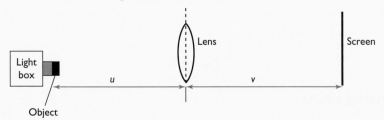

Figure 26

The position of the screen is adjusted until a sharp image appears on the screen. The distance between the object and the screen, u, is measured and the distance between the image and the screen, v, is measured. The procedure is repeated for different values of object distance, u. The results are recorded in the following table.

Object distance, u/m	Image distance, v/m	$\frac{1}{u}$/m^{-1}	$\frac{1}{v}$/m^{-1}
0.50	0.34		
0.45	0.36		
0.40	0.41		
0.35	0.46		
0.30	0.59		

If the lens formula is rearranged and compared to the equation of a straight line we can find out which graph will give us a straight line.

$$\frac{1}{f} = \frac{1}{u} + \frac{1}{v}$$

$$\frac{1}{v} = \frac{1}{f} - \frac{1}{u}$$

$$\frac{1}{v} = -\frac{1}{u} + \frac{1}{f}$$

$$y = mx + c$$

> **Examiner tip**
> Alignment is very important in this experimental set-up. If no optical bench is available a metre rule can be used to line up all the components.

> **Knowledge check 15**
> A student has three lenses, one concave and two convex. He knows the concave lens has a focal length of 25 cm and the convex lenses have focal lengths of 20 cm and 10 cm, but has mixed the lenses up. State the characteristics of the lenses that he can use to identify them.

Content Guidance

Examiner tip

Graphs should use at least half of the grid, have labels and units on both axes, use an easily manageable scale, and have accurately and clearly plotted points, each marked with a small cross or a dot with a circle around it.

Knowledge check 16

Show that the intercept on both axes of the graph of $\frac{1}{v}$ on the y-axis and $\frac{1}{u}$ on the x-axis is $\frac{1}{f}$.

Hence a graph of $\frac{1}{v}$ on the y-axis and $\frac{1}{u}$ on the x-axis will give a straight line with a gradient of –1. The intercept on either axis is equal to $\frac{1}{f}$. Both intercepts are found and an average taken to determine the value of the focal length, f (Figure 27).

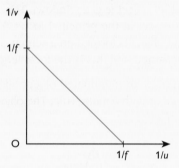

Figure 27

Magnification

The linear magnification, m, of an image is the ratio of the size of the image divided by the size of the object. However, by similar triangles it can be shown that the magnification is also the image distance divided by the object distance (Figure 28).

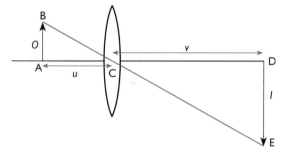

Figure 28

$$\frac{AB}{AC} = \frac{DE}{CD}$$

$$\frac{O}{u} = \frac{I}{v}$$

$$\frac{v}{u} = \frac{I}{O}$$

So

$$\text{magnification } m = \frac{I}{O} = \frac{v}{u}$$

Worked example

An object of length 2 cm is placed 10 cm from a converging lens of focal length 15 cm. Find the length of the image.

> object distance = u = 10 cm
>
> focal length = f = 15 cm
>
> length of object = O = 2 cm

First find the image distance:

$$\frac{1}{f} = \frac{1}{u} + \frac{1}{v}$$

$$\frac{1}{v} = \frac{1}{f} - \frac{1}{u} = \frac{1}{15} - \frac{1}{10} = -\frac{1}{30}$$

> v = −30 cm
>
> magnification $m = \frac{v}{u} = \frac{I}{O}$
>
> $m = \frac{-30}{10} = \frac{I}{2}$
>
> image height = I = −6 cm

This is a magnified virtual image, so this lens may be a magnifying glass.

Power of a lens

The ability of a lens to focus parallel rays of light is called the power of the lens.

> power of lens = $\frac{1}{f}$

The focal length is measured in metres and hence the power is measured in metres^{-1}. 1 metre^{-1} is equivalent to 1 dioptre (1 m^{-1} = 1D).

A converging lens has positive power and a diverging lens has a negative power.

Worked example

A convex lens is placed at the 50 cm mark on a metre rule. Calculate the power of the lens if an image is formed at the 80 cm mark of an object that is placed at the 30 cm mark of the rule. Give your answer in dioptres.

Answer

The answer is to be in dioptres so the distances must be in metres:

> u = 20 cm = 0.2 m
>
> v = 30 cm = 0.3 m
>
> $\frac{1}{f} = \frac{1}{u} + \frac{1}{v}$
>
> $\frac{1}{f} = \frac{1}{0.2} + \frac{1}{0.3}$ = 8.3 D

Examiner tip

When power is to be calculated, convert distances to metres before substituting into the lens formula and do not find the reciprocal at the end unless the focal length is also required.

Knowledge check 17

Compare the power of a lens with its focal length and thickness.

(c) We need to find the object distance that will form a virtual image at the actual far point of 400 cm with this lens.

$$\frac{1}{f} = \frac{1}{u} + \frac{1}{v}$$

$$\frac{1}{30} = \frac{1}{u} + \frac{1}{-400}$$

$$u = 27.9 \text{ cm}$$

So wearing the spectacles to correct the near point will give the person a far point of 28 cm.

(d) A normal far point is at infinity so they want to see an object at infinity and form a virtual image at the actual far point.

$$\frac{1}{f} = \frac{1}{u} + \frac{1}{v}$$

$$\frac{1}{f} = \frac{1}{\text{infinity}} + \frac{1}{-400} = -\frac{1}{400} \text{ cm}^{-1} = -\frac{1}{4} \text{ m}^{-1} = -\frac{1}{4} \text{ D}$$

(e) Use bifocal lenses.

Knowledge check 19

What is the range of distinct vision for a person with normal sight?

Summary

- Lenses refract light. A converging (convex) lens converges parallel rays of light to a real principal focus.

- A diverging (concave) lens appears to diverge parallel rays of light from a virtual principal focus.

- The focal length is the distance between the centre of the lens and the principal focus. The focal length of a lens can be determined approximately using a distant object or, more accurately, using an illuminated object and a screen.

- Ray diagrams can used to diagrammatically find the image formed by a lens. At least two rays must be used to locate the image. Three rays are possible. A ray through the centre of the lens is undeviated. A ray parallel to the principal axis will be refracted through the principal focus of the lens. A ray through the principal focus will be refracted parallel to the principal axis.

- An image is fully described by its position, size (magnified or diminished), orientation (upright or inverted) and nature (real or virtual).

- The lens formula can used to mathematically find the image formed by a lens:

$$\frac{1}{f} = \frac{1}{u} + \frac{1}{v}$$

- The magnification formula is:

$$m = \frac{\text{height of image}}{\text{height of object}} = \frac{v}{u}$$

- The power of a lens is related to the focal length:

$$\text{power} = \frac{1}{f}$$

If the focal length is measured in metres then the unit of power is the dioptre (D).

- The eye refracts light to form sharp images on the light-sensitive retina at the back of the eye. If the eye does not function perfectly, the light may not converge exactly on the retina but before or after it. This means that the image cannot be sharply focused and there is said to be a defect of vision.

- The near point is the nearest point that can be focused by the eye, normally 25 cm. The far point is the furthest point that can be focused by the eye, normally taken to be infinity.

- Myopia is short sight and the light converges before the retina. A diverging lens is used to correct this defect. The focal length is calculated using the lens formula with u taken as the corrected far point and v taken as the virtual uncorrected far point.

- Hypermetropia is long sight and the light converges behind the retina. A converging lens is used to correct this defect. The focal length is calculated using the lens formula with u taken as the corrected near point and v taken as the virtual uncorrected near point.

Superposition and interference

When waves of the same type meet (e.g. two water waves), they pass through each other unaffected.

The principle of **superposition** states that when waves meet the resultant displacement is the vector sum of the individual displacements due to each wave at that point.

If two waves of the same amplitude meet in phase, they add together to give a wave with an amplitude that is the sum of the amplitudes of the original waves. This is called **constructive interference**.

If the two waves of the same amplitude meet out of phase by 180° or π radians (antiphase), the waves cancel each other out. This is called **destructive interference** (Figure 32).

If the waves are not the same amplitude then two waves do not completely cancel each other out.

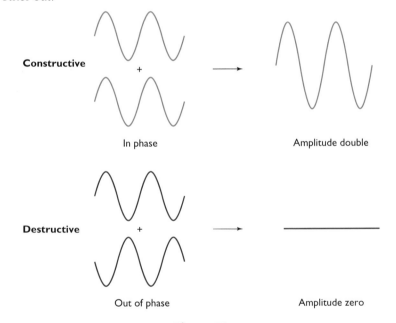

Figure 32

Coherence

Wave sources are coherent if:

- the waves are the same type
- the waves have the same frequency
- the waves always maintain a constant phase difference (the phase difference might be zero but it does not have to be)

and S_2, separated by a distance a. The diffracted light from the two coherent source. overlaps and interferes. An interference pattern of alternate bright and dark fringe: is seen on a screen a distance d from the sources. The fringes are a distance y apart

At a point mid-way between the two sources, the waves have travelled the same distance and are in phase, so constructive interference occurs and a bright fringe i displayed. At points either side of the mid-way point, waves will have different path lengths on arrival at the screen. The waves will give constructive interference if the path difference is a whole number of wavelengths as the waves will then be in phase and a bright fringe will be formed. The waves will give destructive interference if the path difference is an odd number of half-wavelengths as the waves will then be in antiphase and a dark fringe will occur.

This is a useful experiment, as the unknown wavelength of a source of monochromatic light can be determined by measuring the slit separation, the distance of the double slits from the screen and the fringe separation. The wavelength of the light can be calculated using the formula:

$$\lambda = \frac{ay}{d}$$

where λ = the wavelength of the light in metres, a = the separation of the double slit in metres, y = the fringe separation on the screen in metres and d = the distance between the double slit and the screen in metres.

If the double slit was illuminated by white light instead of monochromatic light, the different wavelengths making up the white light would produce their own interference fringe patterns. There would be a central white fringe with dark fringes either side Beyond the centre the maxima and minima of the different colours would overlap and a pattern of coloured fringes would be produced.

Examiner tip

This formula is given on the formula sheet but students often mix up the different quantities. Make sure you know what each symbol represents. Always convert all the units to metres.

Knowledge check 22

Describe the effect on fringe separation if the monochromatic source used in a Young's slit arrangement is replaced with one of longer wavelength.

Worked example

Calculate the wavelength of light that produces fringes of separation 0.95 mm with coherent light from two slits 0.50 mm apart on a screen a distance 0.78 m from the slits. Give you answer in nanometres.

Answer

a = the separation of the slits = 0.50 mm = 0.50×10^{-3} m

y = the fringe separation = 0.95 mm = 0.95×10^{-3} m

d = the distance between the slits and the screen = 0.78 m

$\lambda = \dfrac{ay}{d} = \dfrac{0.5 \times 10^{-3} \times 0.95 \times 10^{-3}}{0.78} = 6.09 \times 10^{-7}$ m = 609 nm

- The principle of superposition states that when waves meet the resultant displacement is the vector sum of the individual displacements due to each wave at that point.
- Constructive interference occurs when two waves that are in phase add together to give a wave with an amplitude that is the sum of the amplitudes of the original waves.
- Destructive interference occurs when two waves of the same amplitude that are in antiphase cancel each other out.
- Superposition of coherent sources causes an interference pattern that consists of alternating maximum and minimum intensities. Constructive interference occurs where waves that meet have travelled the same distance or have a path difference of a whole number of wavelengths, and arrive in phase. Destructive interference occurs where waves that meet have a path difference that is an odd number of half-wavelengths, and they arrive completely out of phase.
- Wave sources are coherent if they are the same type, have the same frequency and maintain a constant phase difference. Observable interference will occur when the waves are coherent and have the same amplitude.

- Standing waves are produced by interference of two waves travelling in opposite directions with the same frequency and amplitude.
- A node is a position on a standing wave with zero amplitude. Nodes are formed at fixed ends. An antinode is a position on a standing wave with maximum amplitude. Antinodes are formed at free ends.
- Standing waves can be formed in a stretched string. A whole number of half-wavelengths are formed in the string. The fundamental frequency is one loop, or half a wavelength, so $f_0 = \frac{v}{2L}$. The overtones are multiples of the fundamental frequency.
- Standing waves can be formed in air pipes closed at one end. An odd number of quarter wavelengths are formed in the pipe. The fundamental frequency is one quarter of a wavelength, so $f_0 = \frac{v}{4L}$. The overtones are odd multiples of the fundamental frequency.
- Young's double-slit experiment demonstrates interference of light using a monochromatic light source. The double slit provides coherent sources that interfere to form bright and dark fringes on a screen. There is a central bright fringe with alternate dark and bright fringes on both sides. The wavelength of the light used can be calculated using the formula:

$$\lambda = \frac{ay}{d}$$

Diffraction

Diffraction is the spreading of waves when they pass through an opening or round an obstacle. The extent of the diffraction depends on the size of the gap compared with the wavelength. Diffraction is most noticeable if the size of the gap is approximately equal to the wavelength of the wave (Figure 48).

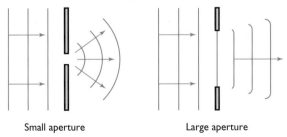

Small aperture   Large aperture

Figure 48

Diffraction explains why you can hear a noise around an open door but cannot see the source. The wavelength of the sound is about the same as the width of the door and so diffracts around it but the wavelength of the light is much smaller than the width of the door.

Light wavelengths are very small so diffraction effects can only be observed when the aperture (gap) is extremely small. When such a narrow slit is used with laser light, a diffraction pattern can be viewed on a screen. There is a central band of maximum intensity and a series of bright and dark bands either side (Figure 49).

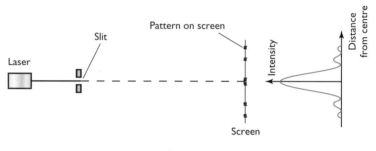

Figure 49

Summary

- Diffraction is the bending or spreading of light through an opening or round an obstacle.

- Diffraction is most noticeable if the size of the gap is approximately equal to the wavelength of the wave.

Sound

Sound is a mechanical, longitudinal wave produced by vibrating objects. As sound propagates it produces regions of high and low pressure known as compressions and rarefactions.

The speed of sound depends on the medium through which it propagates. Sound travels fastest in dense materials such as solids. It travels slower in liquids and slowest in gases. The speed of sound in dry air at room temperature is approximately $340\,\mathrm{m\,s^{-1}}$. Sound cannot travel through a vacuum.

The amplitude of a sound wave corresponds to its intensity or loudness (although intensity and loudness are not exactly the same).

The frequency of a sound wave corresponds to its pitch. Humans can hear frequencies between 20 Hz and 20000 Hz. Frequencies above 20000 Hz are called ultrasonic and frequencies below 20 Hz are called infrasonic.

Experiment: measuring the frequency of sound using a cathode ray oscilloscope

Sound of a single frequency is detected by a microphone connected to a cathode ray oscilloscope (CRO). This sound wave is represented on the CRO as a sine wave, which is a graph of voltage on the vertical scale and time on the horizontal scale. The time base setting controls the horizontal scale. The setting tells us how much time each centimetre of the scale represents. The period of the wave can be determined from the screen and the frequency calculated from the period as the periodic time $T = \dfrac{1}{\text{frequency}}$.

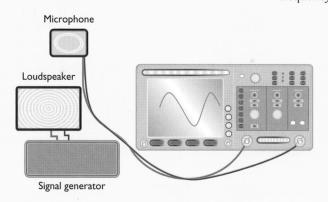

Worked example

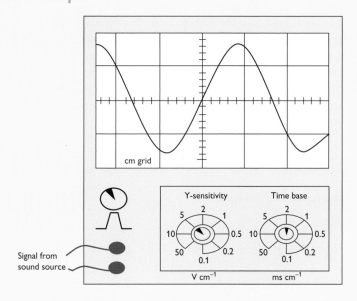

time base = $2\,\text{ms}\,\text{cm}^{-1}$

crest to crest of wave = $3.2\,\text{cm}$

period of wave $T = 3.2 \times 2\,\text{ms} = 6.4\,\text{ms} = 6.4 \times 10^{-3}\,\text{s}$

frequency $f = \dfrac{1}{\text{period}} = \dfrac{1}{T} = \dfrac{1}{6.4 \times 10^{-3}} = 156\,\text{Hz}$

Experiment to measure the speed of sound using a resonance tube

Standing waves in a resonance tube closed at one end can be used to measure the speed of sound (Figure 50).

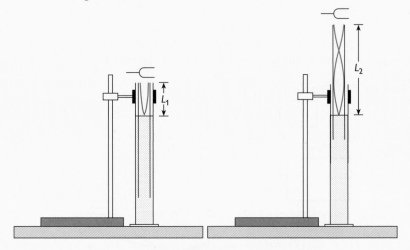

Figure 50

A resonance tube that is open at both ends is placed in a deep tank of water. The water provides a closed end for the tube. Raising and lowering the resonance tube in the water changes the length of the resonance tube. A vibrating tuning fork of known frequency is held over the end of the resonance tube in its lowest position, causing the air to vibrate. The tube is slowly raised increasing the length of the air column until at a certain position the loudest note is heard. This is the first position of resonance. The length L_1 of the air column at this resonant position is recorded. The tube is raised further until a second resonance position is found. This corresponds to the second resonance position. The length L_2 of the air column is also noted.

At the first resonance position

$$L_1 = \frac{\lambda}{4}$$

At the second resonance position

$$L_2 = \frac{3\lambda}{4}$$

$$L_2 - L_1 = \frac{3\lambda}{4} - \frac{\lambda}{4} = \frac{\lambda}{2}$$

So,

$$\lambda = 2(L_2 - L_1)$$

Using the wave equation $v = f\lambda$:

$$v = 2f(L_2 - L_1)$$

where f is the known frequency of the tuning fork in Hz, and L_1 and L_2 have been measured in m. This allows the speed of sound to be calculated in $m\,s^{-1}$.

A second method uses a range of tuning forks of different frequencies to create the standing waves for the first resonance position of each fork. This fundamental node of vibration is a quarter of a wavelength of the sound wave, so $L = \frac{\lambda}{4}$.

$$v = f\lambda \text{ and } \lambda = 4L$$

so,

$$v = 4fL$$

or,

$$L = \frac{v}{4f}$$

Comparing with the equation of a straight line:

$$y = mx + c$$

A graph of L on the y-axis and $\frac{1}{f}$ on the x-axis will be a straight line through the origin with gradient $\frac{v}{4}$.

The speed of sound is approximately $340\,\text{m s}^{-1}$.

Hearing sound

Hearing is the detection of sound by the ear. Hearing depends on certain properties of the sound wave, such as its frequency and intensity. The sound creates variations in air pressure, which the ear converts into electrical signals that are sent to the brain for interpretation (Figure 51).

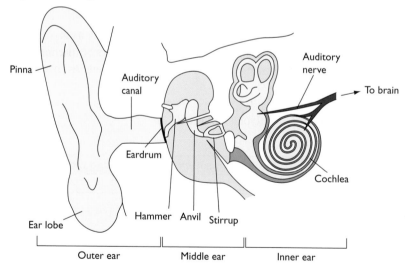

Figure 51

The sound wave starts at the outer part of the ear, which is called the pinna. The pinna plays a minor role in collecting sound waves and directing them into the auditory canal. The auditory canal acts as a pipe, closed at one end by the eardrum. A standing wave will be set up here as resonance occurs at a specific frequency.

The middle ear is an air-filled cavity in which resonance can also occur. It contains three small bones, which provide an efficient mechanical link to transfer the sound power to the inner ear.

The inner ear is a small, liquid-filled cavity containing a spiral tube called the cochlea. Running along the cochlea is membrane composed of hairs of different thickness, length and stiffness. These hairs have different natural frequencies of vibration and so resonate at different frequencies. This is very important, as it is these hairs that provide the mechanism for the ear to distinguish between different sound frequencies. When a sound wave enters the cochlea only certain hairs resonate, causing small electrical signals that are sent to the brain via nerve fibres.

For sounds of frequency below 20 Hz there is no resonant stimulation of the hairs and this gives the lower limit of frequency that can be detected. Above 20 kHz the hairs do not respond to such high frequencies and this gives an upper limit of the detectable frequency range. Hence sound with frequency above 20 kHz is called ultrasound. This upper frequency limit decreases with age as the stiffness of the hairs in the cochlea changes. Some animals can detect frequencies above 20 kHz. Dogs can detect frequencies far into the ultrasound region and hence humans cannot hear a dog-whistle.

The decibel scale

The intensity of sound is the power passing through unit area perpendicular to the wave. Sound intensity is measured in watts per square metre ($W\,m^{-2}$). The ear can detect a wide range of sound intensities from the lowest intensity, called the threshold of hearing, at $10^{-12}\,W\,m^{-2}$ to an upper limit of approximately $100\,W\,m^{-2}$. At the upper limit these intensities are painful and can result in permanent damage to the ear.

Detection of sound by the ear is not linear. Small changes in intensity are more easily distinguished at low intensities than at higher intensities. The response of the ear to different intensities is logarithmic.

Examiner tip
Learn that the threshold of hearing is at
$1.0 \times 10^{-12}\,W\,m^{-2}$.

The intensity level is the intensity relative to an agreed zero of intensity, which is taken to be the threshold of hearing: $1.0 \times 10^{-12}\,W\,m^{-2}$ at frequencies of about 2 kHz.

$$\text{Intensity level} = 10\log_{10}\frac{I}{I_0}$$

Intensity level is measured in decibels, dB (Table 3).

Table 3 Intensity levels of typical sounds

Sound	Sound intensity/$W\,m^{-2}$	Sound intensity level/dB
Threshold of hearing	10^{-12}	0
Whispering	10^{-9}	30
Speech	10^{-5}	70
Factory noise	10^{-3}	90
Rock music	10^{-1}	110
Noise causing discomfort	10^{0}	120
Threshold of pain	10^{1}	130

Examiner tip
Learn the graph of intensity response with frequency for the average ear and be able to explain the main features.

Graph of frequency and intensity response for the ear

The ear responds to sound using resonance. The vibrations of the sound match the natural frequency of certain parts of the ear. The brain discriminates frequencies by identifying the different resonating parts of the ear. As there are several parts of the ear that resonate, the required intensity of a sound wave incident on the ear, such that a particular frequency can be detected, is not constant. Resonance in the auditory canal means that lower intensities can be heard for frequencies around 2–3 kHz (Figure 52).

The range of frequencies detectable by the human ear is from about 20 Hz to 20 kHz at normal, everyday intensity levels. This range depends on the intensity of the sound. Higher intensities give rise to a larger range of audible frequencies. However, this is limited by the discomfort and possible ear damage that can be caused by high intensities. The ability to discriminate between frequencies varies. From 60 Hz to 1 kHz a difference of 3 Hz can be detected but this ability diminishes as the frequency increases, and above 10 kHz is poor.

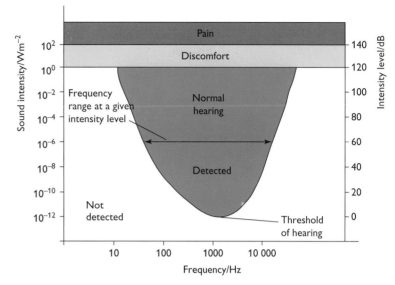

Figure 52

Figure 52 shows that the threshold of hearing is $1.0 \times 10^{-12}\,\mathrm{W\,m^{-2}}$, which corresponds to an intensity level of 0 dB. The ear is most sensitive at approximately 2000–3000 Hz and hence it is only at this frequency that the threshold intensity is detectable. The width of the curve indicates the detectable frequency range at a particular intensity.

Worked example

(a) What is the range of frequencies detectable by a typical ear?

(b) What is meant by the threshold of hearing?

(c) (i) Describe how the sensitivity of the ear changes across the detectable range of frequencies.
   (ii) Sketch a graph to show how the minimum detectable intensity varies with frequency for a typical ear.
   (iii) Mark on the graph the frequency at which the ear is most sensitive.
   (iv) Shade the part of the graph where sounds are not detectable

(d) A factory worker is exposed to a working environment where there is a constant sound intensity of $0.52\,\mathrm{W\,m^{-2}}$.
   (i) What sound intensity level is the worker exposed to in the factory?
   (ii) What will exposure to this intensity level do to his hearing and how can he protect against this?

Answer

(a) 20 Hz–20 kHz

(b) The threshold of hearing is the minimum sound intensity detectable at a particular frequency.

(c) (i) The sensitivity increases from 20 Hz to a maximum at 2 kHz and then decreases again to 20 kHz.

(ii)–(iv)

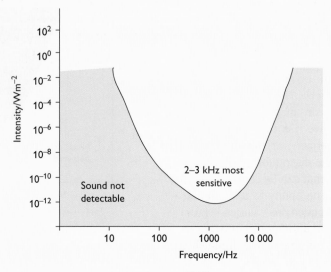

(d) (i) Intensity level $= 10\log_{10}\frac{I}{I_0} = 10\log_{10}\frac{0.52}{10^{-12}} = 117\,\text{dB}$

(ii) The intensity level may cause permanent damage to his hearing unless he wears ear defenders.

Summary

- The frequency of a pure note, for example from a tuning fork or a signal generator, can be measured using a microphone and a cathode ray oscilloscope. The time base setting allows the period of the wave to be determined. The frequency is found from $f = \frac{1}{T}$.

- The speed of sound can be determined using a resonance tube that is closed at one end. A standing wave is set up in the resonance tube at the shortest length using a sound of known frequency. At resonance the loudest sound is heard. This wave is a quarter of a wavelength. This is repeated at the next position of resonance, which is three-quarters of a wavelength. From this the wavelength can be determined and with the known frequency and the wave equation, the velocity can be found.

- The detection of sound by the ear is known as hearing and this depends on certain properties of the sound wave, such as its frequency and intensity.

- The intensity of sound is the power passing through unit area perpendicular to the wave. It is measured in watts per square metre (Wm^{-2}). The lowest intensity detectable is called the threshold of hearing and is $10^{-12}\,\text{Wm}^{-2}$. The response of the ear to different intensities is not linear — it is logarithmic.

- Intensity level is measured in decibels dB and can be calculated using:

  intensity level $= 10\log_{10}\frac{I}{I_0}$

  (The threshold of hearing is

  $I_0 = 1.0 \times 10^{-12}\,\text{Wm}^{-2}$

  which corresponds to an intensity level of 0 dB.)

- The ear responds to sound using the differing resonant frequencies of parts of the ear. Humans can detect frequencies of 20 Hz–20 kHz at normal everyday intensity levels. A graph can be drawn to indicate that the ear is most sensitive at approximately 2000–3000 Hz and hence it is only at this frequency that the threshold intensity is detectable. The width of the curve indicates the detectable frequency range at a particular intensity.

Imaging techniques

Imaging techniques are non-invasive procedures to enable medical staff to examine organs and structures inside the human body without the need for exploratory surgery. Using imaging techniques means that patients do not require a long recovery time, which they would after surgery and anaesthetic, and the risk of infection is reduced. These techniques can be deployed in outpatient departments and so avoid the cost of using hospital beds.

Flexible endoscopes

Endoscopes (Figure 53) are used to view internal parts of the body and to perform minor operations. Optical fibres are used in endoscopes. They are narrow tubes of glass fibres with a plastic coating that carry light from one end to the other. The light bounces off the walls of the fibre as the light rays undergo total internal reflection to be trapped in the fibres. In order for this to be achieved, the light ray must be incident on the walls of the fibre at an angle greater than the critical angle for light travelling from glass to plastic.

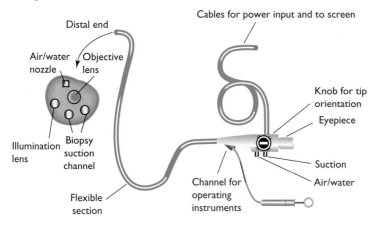

Figure 53

The flexible shaft

The shaft of the endoscope is only 10 mm in diameter and can be up to 2 m long. It is flexible to make it easy to manoeuvre through the body and is coated in steel and plastic in order to make it waterproof and resistant to chemical damage.

The flexible shaft includes:
- a non-coherent optic bundle
  - Light is guided to the area under investigation by a non-coherent fibre optic bundle in which the optical fibres are randomly aligned.
- a coherent optic bundle
  - The image must be transmitted back to the observer by a coherent fibre optic bundle of ordered, parallel fibres, which are lined up at both ends so that an image can be transmitted (Figure 54). In order to produce a clear image, the shaft contains up to 10 000 fibres.

Knowledge check 24

Explain why it is advantageous to patients that medical imaging techniques allow doctors to diagnose illness without resorting to surgery.

Knowledge check 25

Explain why a coherent bundle of fibres is used to transmit the image out of the body but non-coherent fibres are used to transmit light in to illuminate the internal structure.

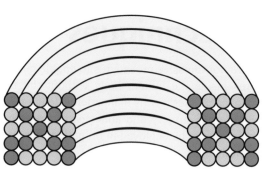

Figure 54

- a distal end
  - This is inserted into the patient's body and is able to be bent in the desired direction. The image is focused by an objective lens on the end.
- a water pipe
  - Water is carried to the objective lens to wash it and keep the view clear.
- operations channel
  - Tools are deployed at the distal end for surgery, for example a laser to cut tissue or seal a wound.

Uses of endoscopes

The endoscope can be inserted into any natural opening of the body, allowing internal structures such as the oesophagus, stomach or colon to be examined. Often a small incision is made near the structure to be examined and the endoscope inserted.

- Bronchoscopy — the endoscope is inserted through bronchial tubes within the lungs in order to look at the airway and to remove any objects causing a blockage.
- Gastroscopy — the endoscope is inserted down the throat to look for problems with the oesophagus, stomach and duodenum, such as bleeding or ulcers.
- Laparoscopy — the endoscope is inserted through an incision in the abdominal wall in order to look at abdominal organs and perform minor surgery.

Ultrasound

The reflection of ultrasound signals at tissue boundaries is the basis of ultrasound diagnosis.

Ultrasound is sound with frequencies greater than 20 000 Hz, which is higher than the human ear can detect. Ultrasound imaging uses high-frequency sound waves to image the internal structures in the body. High frequencies of ultrasound produce better resolution. On the other hand, higher frequencies of ultrasound have short wavelengths and are absorbed easily and therefore are not as penetrating. For this reason, high frequencies are used for scanning areas of the body close to the surface and low frequencies are used for areas that are deeper down in the body. These frequencies generally have a range of 1–15 MHz, with a typical power of 0.1 mW.

How ultrasound is produced and detected

Ultrasound is produced and detected using an ultrasound transducer. Ultrasound transducers convert electrical energy into sound energy and hence transmits an

Examiner tip

It is important to appreciate the wide range of functions that the endoscope can carry out in addition to imaging. The tool aperture can carry a laser or a surgical implement such as a scalpel.

Examiner tip

Remember that ultrasound is just like all other sounds in that it is a longitudinal wave made by vibrations. The difference is the very high frequency.

ultrasound signal into the body to reflect at the boundary between two biological structures. The same transducer can detect the echoed sound and convert it to an electrical signal to be analysed (Figure 55).

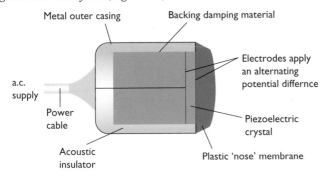

Figure 55

- The piezoelectric crystal changes its dimension when a potential difference is applied across it.
- The acoustic insulator prevents any external sound sources affecting the crystal.
- The backing material prevents the continuation of crystal oscillations when the a.c. voltage is removed.
- The plastic membrane vibrates to transmit and receive the ultrasound.
- The co-axial cable provides the source of a.c. voltage.

To produce an ultrasound, a piezoelectric crystal has an alternating current applied across it, causing it to expand and contract at the supply frequency. This conversion of electrical energy to mechanical energy is known as the piezoelectric effect. The mechanical vibrations create sound, which is transmitted into the body and then bounces back off the object under investigation. The sound hits the piezoelectric crystal and then has the reverse effect — the echoed sound is converted to mechanical energy as it vibrates the crystal and this is converted into electrical energy. By measuring the time between when the sound was sent and when the echo was received, distances can be calculated. As the transducer acts as an emitter and a receiver, it emits only a very short pulse of ultrasound separated by a short gap, during which it is in receive mode.

Note: Ultrasound waves are strongly reflected at the air–skin boundary. To overcome this problem a water-based cellulose jelly is smeared on the skin. This jelly acts as a coupling agent to ensure that most of the ultrasound enters the body. Ultrasound is not used to image the lungs or bowel due to the presence of air and the strong reflection that would occur.

A-scans and B-scans

A-scans (amplitude scans) can be used to measure distances or ranges in the body (Figure 56). A transducer emits an ultrasonic pulse and the time taken for the pulse to bounce off structures in the body and come back is plotted on the horizontal axis of a CRO using the time base against the amplitude of the reflected pulse on the vertical axis. The ultrasound signal will become weaker as it travels further into the body and hence signals from deeper surfaces are amplified. A-scans only give one-dimensional

Examiner tip

Transducers convert energy from one form to another. Remember that here the transducer converts electrical energy to sound energy when in transmission mode and sound energy to electrical energy when in detection mode.

Knowledge check 26

Explain the importance of the acoustic insulator in the ultrasonic transducer.

information and therefore are not useful for forming images. A-scans are used to measure foetal head diameter and eyeball depths.

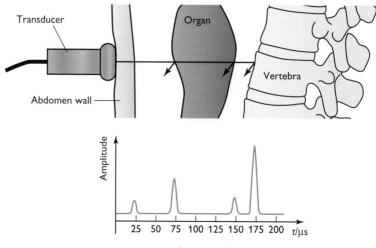

Figure 56

B-scans (brightness scans) use the same principle of reflected ultrasound pulses but can take an image of a cross-section through the body (Figure 57). The transducer is swept across the area to be examined at different angles and the time taken for pulses to return is used to determine distances, which are mapped as a series of dots. The strength of the reflected pulses is used to determine the brightness of the dots. The data are stored and used to produce an image on a monitor. B-scans give two-dimensional information about the cross-section. They can be used to identify tumours, but the most common use is to monitor foetal development.

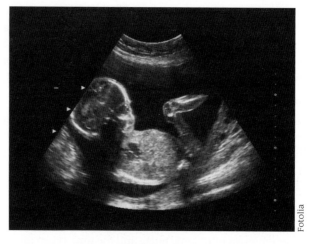

Figure 57

Dangers of ultrasound

Ultrasound waves cause a heating effect in tissue. They can also cause large-amplitude resonant vibrations. This effect can be utilised to break up kidney stones.

Worked example

(a) Name a type of transducer used to produce ultrasound.

(b) Figure 58 shows a cross-section of a human body through which an ultrasound signal passes during an A-scan examination.

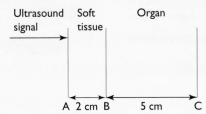

Figure 58

(i) The speed of ultrasound in soft tissue is $1500\,ms^{-1}$. Find the time interval between a pulse being sent out and its echo being received from surface B.

(ii) The pulse returns from surface C after a time interval of $65\,\mu s$. Calculate the speed of ultrasound in the organ.

Answer

(a) piezoelectric crystal

(b) (i) $\text{time} = \dfrac{\text{distance}}{\text{speed}}$

The round trip distance for the ultrasound is $2 \times 2\,cm = 4 \times 10^{-2}\,m$.

$\text{time interval} = \dfrac{4 \times 10^{-2}}{1500} = 2.7 \times 10^{-5}\,s = 27\,\mu s$

(ii) time in organ = total time − time in soft tissue

$= 65 \times 10^{-6}\,s - 27 \times 10^{-6}\,s = 38 \times 10^{-6}\,s$

distance travelled in organ = 2 × thickness of organ

$= 2 \times 5\,cm = 10\,cm = 10 \times 10^{-2}\,m$

$\text{speed} = \dfrac{\text{distance}}{\text{time}} = \dfrac{10 \times 10^{-2}}{38 \times 10^{-6}} = 2600\,ms^{-1}$

X-rays in medicine: CT scanning

X-rays can be used to examine internal structures in the body because the various types of body tissue absorb different amounts of X-ray radiation. X-rays penetrate soft tissue but are stopped by bone. Photographic film is sensitive to X-ray exposure and a shadow image of the bone is obtained.

Computed tomography (CT) imaging, also known as 'CAT' (computed axial tomography) scanning, uses X-ray technology to create detailed, three-dimensional images.

X-rays are a form of ionising electromagnetic radiation and can therefore cause damage to living cells (especially to the DNA) as some of the energy of the X-rays will be absorbed by the body tissue. X-rays have very high frequencies and very short wavelengths. Their wavelengths range between 0.001 nm and 10 nm.

The X-ray tube

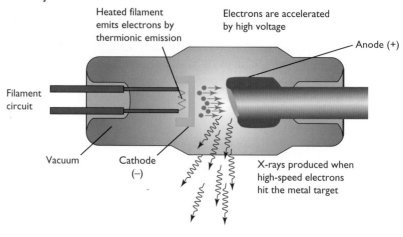

Figure 59

Electrons are emitted by thermionic emission from the heated, positively charged filament and accelerated towards the positively charged tungsten target (Figure 59). The X-ray tube is evacuated to ensure the electrons do not collide with any gas particles and deviate from their path. The tungsten target absorbs the electrons and releases some of the energy in the form of X-rays by one of two methods:

- The rapid deceleration of electrons after passing the nucleus.
- The tightly bound inner electrons being knocked out of atoms by the incident electrons and higher-state electrons dropping down to fill the vacancy.

This process is inefficient because there is a large amount of energy released as heat. For this reason the tungsten target has a copper mounting to conduct heat and it is cooled by circulating oil through the mount. Spinning the tungsten target at high speed also helps to stop it overheating. Narrower beams of X-rays produce a sharper image. The tungsten target is therefore angled so that a wide beam of electrons produces a narrow beam of X-rays

Using X-rays to form images

- Hard X-rays are X-rays with a higher frequency and are more penetrating than soft X-rays. Soft X-rays are usually filtered out when doing a scan because they cannot penetrate through a patient's body and add needless risk of radiation damage.
- Attenuation is a measure of how much something absorbs X-rays. The amount of attenuation increases with the number of protons in the nuclei. For example, bones have a higher attenuation than soft tissue and therefore produce a dark shadow when X-rayed, whereas soft tissue appears much fainter.
- When patients have an X-ray, they are usually scanned at a frequency of approximately 7×10^8 Hz because body tissues absorb this frequency the best.
- X-rays are best suited to imaging bones and have a very high resolution. For imaging soft tissue, however, there is very little contrast and so a contrast medium is needed. Contrast mediums are substances given to the patient that absorb X-rays and produce an image of the area under investigation when X-rayed.

lanar, conventional X-ray images are always two-dimensional projections of three-dimensional objects. In addition, there is always the possible problem of one object iding another. One way to overcome these problems is to image successive two-dimensional slices of the patient to build up a three-dimensional image. This is called omography, from the Greek word *tomos*, meaning a section.

omputed tomography (CT) scans are better than conventional X-rays for imaging oft tissue. In fact, CT has the unique ability to image a combination of bone, soft issue and blood vessels. CT requires the use of powerful digital computing to rocess multiple images of slices of the body taken by a rotating X-ray device from ifferent angles. A number of detectors around the patient capture the transmitted -ray images and the three-dimensional image is built up from the combined data. he tube voltages and exposures are greater than for conventional X-radiography.

Knowledge check 27

Explain how X-rays are produced when accelerated electrons hit a metal target.

Comparison of CT with conventional X-rays

CT images are three-dimensional and therefore more detailed.
CT can distinguish similar tissues types, but conventional X-ray can only distinguish bone and soft tissue.
CT involves a much greater dose of ionising radiation.
CT requires complex computational capability.
In CT scans the X-ray tube moves and it is prone to breaking.

Comparison of CT and ultrasound

Both produce slices through body.
CT and X-ray detect transmitted signals while ultrasound detects reflected signals.
CT uses ionising radiation, while it is generally thought that ultrasound is safe, hence its use in obstetrics.
Resolution with ultrasound is better than with CT due to the longer exposures needed for CT.
Ultrasound does not pass through air due to extreme reflections, so cannot be used to study lungs.
Ultrasound cannot penetrate bone, so CT is used for brain examinations.

Magnetic resonance imaging

Magnetic resonance imaging (MRI) provides detailed images for use in medical iagnosis. Radio frequency radiation is used, so this form of imaging is considered to arry no risk to the patient, but the process is noisy and claustrophobic.

he image is produced by analysing radio-frequency signals emitted by certain atoms n the body after they have been made to resonate. The basis of this technique is uclear magnetic resonance (NMR):

Nuclei with unequal numbers of protons and neutrons spin.
Spinning nuclei act like tiny magnets, but are randomly orientated.
They will align themselves with a powerful external magnetic field (0.1–4 tesla) provided by super-conducting magnets.

- Just like a spinning top that is coming to the end of its motion, the spinning nuclei do not spin precisely around one point. There is a wobble in the spin, causing them to precess around the direction of the field.
- The frequency of precession depends on the applied magnetic field and the nucleus, and is called the Larmor frequency.
- Applying a pulse of radio-frequency electromagnetic radiation at this Larmor frequency causes resonant absorption of energy and the nuclei flip their spin and move to a higher energy state.
- When the radio pulse ends, the nuclei return to the equilibrium lower-energy state with the emission of radio electromagnetic radiation with a characteristic relaxation time.
- Different relaxation times result in varying bright and dark spots on the image.
- The re-emitted radio waves and the relaxation times are used with advanced computational techniques to form an image.
- In addition to the main static magnetic field, there is a second set of coils called field gradient coils, which allow variations of field along the length of the patient. Hence, different parts of the body have different Larmor frequencies, allowing small parts of the body to be isolated and examined in detail.
- Hydrogen nuclei are used as the imaging agent due the abundance of hydrogen in human tissues.
- MRI is very successful in imaging the brain and spinal cord.

Strong magnetic fields are needed and can be supplied by superconducting magnets, which require cooling with liquid helium ($-269°C$); this is expensive. Before the MRI scanning process can begin, patients must remove all metal objects, such as jewellery or watches, because they can interfere with the magnetic field. Once this has been done, the patient is instructed to lie on a bed and is placed into a magnetic field. The radio waves that are re-emitted by the nuclei in the patient's body are detected by sensors, which are placed around the body. Next, a computer assembles the slices into a three-dimensional image, allowing doctors to make a complete diagnosis. The scan can take as long as 30 minutes.

Hazards

- Large magnetic fields cause the body to be magnetised and this causes stress on the body.
- Patients with metal implants or pacemakers are excluded from this imaging technique.
- Special sealed rooms in hospitals are required to house magnetic equipment.

Summary

- Imaging techniques are non-invasive medical procedures that enable medical staff to examine internal organs without the need for exploratory surgery.

- Endoscopes use optical fibres and total internal reflection to view internal parts of the body. The flexible shaft contains a randomly arranged non-coherent fibre bundle to illuminate the area under investigation and an ordered coherent bundle of multiple parallel fibres to transmit the image back to the observer. A water channel and a tools attachment are also at the distal end. The endoscope can be inserted into any natural opening of the body. For example, it can be fed down the throat to look for problems with the oesophagus, stomach and duodenum, such as bleeding or ulcers.

- Ultrasonic imaging uses ultrasound reflections to image the internal structures in the body. The frequencies generally range from 1–15 MHz, with a typical power of 0.1 mW. Ultrasound is produced and detected using an ultrasound transducer, which converts electrical energy into sound energy. The ultrasound signal is reflected at the boundary between two structures. To overcome excessive reflection at the air–skin boundary, a water-based cellulose jelly is smeared on the skin. Ultrasound is not used to image the lungs or bowel due to the presence of air and the strong reflection that would occur. Amplitude scans measure distances in the body. Brightness scans produce two-dimensional images.

- Computed tomography (CT) scans use X-rays to image the body. X-rays penetrate soft tissue but are stopped by bone. Photographic film is sensitive to X-ray exposure and a shadow image of the bone is obtained. X-rays are ionising radiation so the body tissue absorbs some of the energy of the X-rays and the living cells are damaged. An X-ray tube produces X-rays by bombarding a metal target with accelerated electrons. Conventional X-rays are used to produce planar two-dimensional images. CT scans use a moving X-ray tube and a range of detectors to produce slices of the body, which are computed to form a three-dimensional image. The body receives more ionising radiation in a CT scan than in a conventional X-ray.

- Magnetic resonance imaging (MRI) uses radio-frequency radiation to resonate spinning hydrogen nuclei that have been subjected to a powerful external magnetic field (0.1–4 tesla) provided by superconducting magnets. When the radio pulse is removed the nuclei relax in a characteristic time. Different relaxation times results in varying bright and dark spots on the image. The scanner builds up information from many slices through the body and a computer assembles the slices into a three-dimensional image. MRI is very successful at imaging the brain and spinal cord.

Photon model

The photoelectric effect

In a metal each atom has a few loosely attached outer electrons, which move randomly through the material as a whole. If one of these electrons near to the surface of the metal tries to escape, it experiences an attractive inward force from the resultant positive charge left behind. An electron cannot escape the metal unless an external source does work against this attractive force and increases the kinetic energy of the electron. This can be achieved by heating (thermionic emission) or by shining electromagnetic radiation on the metal (photoelectric emission).

Photoelectric emission is the release of electrons from a metal when electromagnetic radiation of high enough frequency is incident on its surface. The electrons emitted are called photoelectrons.

The photoelectric effect can be demonstrated using a gold leaf electroscope. Ultraviolet radiation is directed onto a small sheet of zinc freshly cleaned with emery cloth to remove any oxidation, and connected to an electroscope.

Examiner tip
Classical physics demonstrated and explained how light waves can be reflected, refracted, diffracted, polarised and superposed. Light was well understood before the photoelectric effect was demonstrated.

When the electroscope is charged positively the ultraviolet radiation has no effect on the gold leaf.

When the electroscope is charged negatively the ultraviolet radiation causes its immediate discharge, indicated by the gold leaf returning to its vertical position (Figure 60). A sheet of glass between the ultraviolet and the zinc halts the discharge.

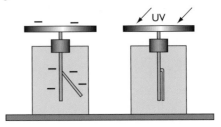

Figure 60

Results of the photoelectric effect

The number of photoelectrons emitted per second depends on the intensity of the incident radiation. This is justifiable in that the more intense the radiation the more energy is absorbed by the metal and the more electrons are able to escape.

The photoelectrons are emitted with a range of kinetic energies from zero up to a maximum value, which increases with the frequency of incident radiation and is independent of intensity. This is a surprise in that classical wave theory would expect photoelectrons to have greater kinetic energy if the intensity of the radiation was greater. For each metal, there is a certain threshold frequency below which no photoelectric emission occurs, no matter how intense the radiation. Also, photoemission occurs immediately. There is no time delay while the electrons build up the energy required for emission. Classical wave theory cannot explain these observations.

The **work function** ϕ of a metal is the energy that must be supplied to enable an electron to escape from its surface.

ϕ is often quoted in units of electronvolts. The electronvolt (eV) is an equivalent unit of energy. It is equal to the kinetic energy gained by an electron in being accelerated by a potential difference of 1 volt. 1 eV is equivalent to 1.6×10^{-19} J. It is a much more convenient unit of energy than the joule when discussing atoms.

Worked example

The energy required to ionise an argon atom is 15.8 eV. Express this in joules.

Answer

$$1\,eV = 1.6 \times 10^{-19}\,J$$

$$15.8\,eV = 15.8 \times 1.6 \times 10^{-19} = 25.3 \times 10^{-19}\,J$$

Photons

Planck tried to solve the inadequacies of the wave theory. He stated that rather than electromagnetic radiation being emitted continuously it is emitted in discrete amounts. He said that energy is quantised; it has specific values with no values in between.

A quantum of energy is a packet of energy. Energy is only considered in whole numbers of packets (quanta). Each quantum contains energy dependent on the frequency of the electromagnetic radiation.

The energy, E, of a quantum is given by:

$$E = hf$$

where h is Planck's constant ($h = 6.63 \times 10^{-34}$ Js) and f is the frequency of the electromagnetic radiation in hertz.

From wave theory, the speed of a wave is given by $v = f\lambda$. All electromagnetic waves travel at the speed of light. c. Hence for electromagnetic waves we can write $c = f\lambda$.

So,

$$E = hf = \frac{hc}{\lambda}$$

Worked example

Calculate the energy of quanta of red light of wavelength 656.3 nm.

Answer

$$E = hf = \frac{hc}{\lambda} = \frac{6.63 \times 10^{-34} \times 3 \times 10^8}{656.3 \times 10^{-9}}$$

$$E = 3.03 \times 10^{-19}\,\text{J} = \frac{3.03 \times 10^{-19}}{1.6 \times 10^{-19}}\,\text{eV} = 1.89\,\text{eV}$$

Einstein extended Planck's ideas in 1905 by deriving an equation that explained the laws of photoelectric emission. He assumed that electromagnetic radiation was not only emitted in whole numbers of quanta but that they travelled in quanta and were absorbed in quanta. He called a quantum of electromagnetic energy a photon (Figure 61).

When a photon interacts with an electron, it transfers all its energy to a single electron instantaneously. If the frequency of incident radiation is below the threshold frequency for that metal, the energy carried by the photon is insufficient for an electron to escape. If the frequency of radiation is equal to the threshold frequency, the energy carried by each photon is just sufficient for the electrons at the surface to escape.

If the energy of the photon hf is greater than the threshold energy ϕ, the electron escapes and the excess energy appears as kinetic energy of the photoelectron.

$$hf - \phi = \frac{1}{2}mv_{max}^2$$

This is Einstein's photoelectric equation.

Knowledge check 28

Explain what the term quantised means.

However, an atom in an excited state can also be stimulated to do this. A photon of exactly the right frequency can induce the excited atom to emit a photon of the same frequency and in phase with the stimulating photon and travelling in the same direction. These two identical photons can then stimulate more identical photons to be emitted and the chain reaction continues. The number of identical photons increases — it is amplified. To produce many of these identical photons, many excited atoms are needed. Atoms are pumped into excited states and remain in these states for a longer time than normal. Such a condition is called an inverted population. Particular materials that give these states are used in a laser. An example is synthetic ruby.

Uses of lasers

Lasers have a wide range of practical applications. Some examples are listed below:
- Keyhole surgery and eye surgery — lasers make precision instruments.
- Fibre-optic communication — lasers transfer huge amounts of information at the fastest speeds.
- Barcode scanning — products have unique identity numbers read quickly by lasers.
- Reading DVDs and CDs — lasers retrieve data by reflecting off microscopic pits on the discs.

Characteristics of laser light

- Monochromatic — photons have the same energy and hence frequency.
- Coherent light — photons are all in phase.
- Intense light — photons are:
  - in phase, so constructive superposition gives high amplitude
  - collimated, as all the photons travel in the same direction

Summary

- The energies of the electrons in an atom can only have certain values, called energy levels. All atoms of a given element have the same set of energy levels and these are characteristic of the element.

- The energy levels of an atom are usually represented as a series of horizontal lines.

- The lowest energy state of an atom is its ground state. An atom can absorb energy and promote an electron to a higher energy level. The atom is now unstable and is in an excited state.

- The electron will fall back to a lower energy level with the emission of a photon of energy: $E_2 - E_1 = hf$. The photons form an emission line spectrum with bright lines at specific frequencies on a dark background.

- The levels have negative values because the energy must be supplied to free an electron and ionise the atom.

- Laser is an acronym for light amplification by stimulated emission of radiation.

- An electron in an excited state can be stimulated to fall to a lower energy level. A stimulating photon of exactly the right frequency can induce this fall, and a photon of the same frequency, in phase and travelling in the same direction is emitted. These two identical photons can then stimulate more identical photons to be emitted and the chain reaction continues.

- Laser light is monochromatic, coherent, intense and collimated.

- Lasers can be used in fibre optic communication, barcode scanning and reading DVDs and CDs.

Wave–particle duality

Many phenomena are fully explained by the wave theory of light. For example, diffraction, interference and polarisation are evidence that the wave nature of light is a valid theory. However, the photoelectric effect requires another explanation of the behaviour of light. Planck and Einstein declared that light was a stream of particles called photons — quanta of energy. Therefore light exhibits wave–particle duality — it has properties of both waves and particles.

De Broglie predicted that matter also exhibits wave–particle duality. He suggested that a particle with momentum p has an associated wavelength λ given by the formula:

$$\lambda = \frac{h}{p}$$

where λ is the de Broglie wavelength in m, p is the momentum of the particle in $kg\,m\,s^{-1}$ and h is Planck's constant.

Worked example

What is the de Broglie wavelength of an electron accelerated through a potential of 1000 V?

Answer

The kinetic energy gained by the electron will be equal to the charge on the electron multiplied by the potential.

$$\tfrac{1}{2}mv^2 = eV = 1.6 \times 10^{-19} \times 1000 = 1.6 \times 10^{-16}$$

So

$$v = \left(\frac{2 \times 1.6 \times 10^{-16}}{9.1 \times 10^{-31}} \right)^{\frac{1}{2}} = 1.9 \times 10^{7}\,m\,s^{-1}$$

momentum $p = mv$

So

$$\lambda = \frac{h}{p} = \frac{h}{9.1 \times 10^{-31} \times 1.9 \times 10^{7}} = 3.9 \times 10^{-11}\,m$$

Electron diffraction

The electron wavelength calculated in the above worked example is comparable to the wavelength of X-rays. X-rays can be diffracted because they are waves, so electron diffraction would confirm particles exhibiting wave-like behaviour. For observable diffraction the space through which the wave passes must be of the order of the wavelength of the wave. X-rays are diffracted by the spacing between crystals, so electrons must be diffracted by the same structures.

The wave nature of particles was confirmed by electron diffraction (Figure 65).

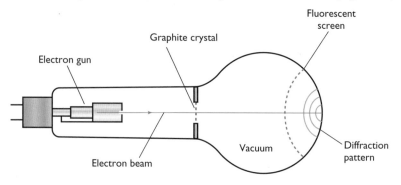

Figure 65

A crystal of graphite was used to diffract the accelerated electrons and an interference pattern was produced similar to the concentric rings produced by X-ray diffraction. If the accelerating voltage was increased the spacing of the rings decreased, showing that the electrons had a shorter wavelength.

Worked example

Calculate the de Broglie wavelength of a 60 g ball travelling at $40\,\text{m s}^{-1}$. Comment on the magnitude of this wavelength.

Answer

$$\lambda = \frac{h}{p} = \frac{6.63 \times 10^{-34}}{0.06 \times 40} = 2.8 \times 10^{-34}\,\text{m}$$

Compared with the mass of an electron, this ball is massive and therefore its wavelength is small to produce an observable wave-like effect. Hence objects like this and larger are characterised by their particle nature.

Summary

- Wave experiments such as Young's slits, diffraction or polarisation demonstrate the wave nature of light. The photoelectric effect demonstrates the particle nature of light. So the behaviour of light can be explained as a wave or as particles. This dual nature is called wave–particle duality.

- Electrons exhibit wave–particle duality. Electrons behave as particles when they collide.

- Electrons also behave as waves because they can be diffracted by a thin sheet of graphite. For observable diffraction the wavelength should be approximately equal to the size of the gap. The atomic spacing of the graphite is used as the gap for electron diffraction. The resulting diffraction pattern is a series of concentric circles. The radius of the circles decreases with increasing speed of the electrons.

- De Broglie stated that the wavelength of a moving particle with momentum p is given by $\lambda = \frac{h}{p}$.

CCEA AS Physics

Questions & Answers

The unit assessment

Unit AS 2 is a written examination of duration 1 hour 30 minutes. It consists of a number of compulsory, short, structured questions. Some of the questions may require an extended response of several sentences. The exam is designed to assess your understanding of all elements in the specification for this unit and all questions must be attempted. It is therefore essential that you revise all sections of the unit.

The exam incorporates an assessment of the **quality of written communication**. Make sure your responses are legible and that spelling, punctuation and grammar are accurate. Use well-structured sentences starting with a capital letter and ending with a full stop. Present information clearly, in a logical sequence, and use appropriate scientific language.

Some questions will require you to demonstrate your knowledge and understanding of physics, and some questions will require you to apply this understanding to unfamiliar situations. It is important to remember that when presented with an unfamiliar situation, the principles of physics are the same and you have all the tools at your disposal to solve the problem. Be confident in your approach to the questions.

The examination will also involve the assessment of 'How Science Works' — a consideration of how scientific knowledge is developed, interpreted, evaluated and communicated. For example, this element may be included in a question about experimental techniques.

Command terms

Examiners use certain words that indicate the type of response required. It is helpful to be familiar with these terms.

- **State** — an exact, concise statement in words.
- **Define** — a word equation or a symbol equation with all the terms defined.
- **Explain** — an extended answer using correct physics terminology. The depth of the answer should reflect the number of marks available.
- **Describe** — a fuller answer, which may be enhanced with an appropriate diagram.
- **Calculate** — a numerical answer, showing all working. The number of significant figures in the answer should be consistent with the data, but each stage of the calculation should be kept in full in your calculator to avoid excessive rounding.
- **Determine** — the value cannot be obtained directly but some data may be extracted from another source, e.g. a graph, and used to obtain the answer.
- **Show** — a value is given and you must perform a calculation, showing all your working to lead to this value. The value should not be used in the calculation. Give your answer to more significant figures than the given value to prove you have done the calculation.
- **List** — a series of words or terms, possibly in a specific order.

- **Sketch** — usually a graph showing a specific trend, with the axes labelled, including the origin if appropriate, and a scale if numerical data are given.
- **Draw** — a carefully drawn diagram, which is fully labelled and includes all available measurements.
- **Estimate** — a calculation involving a reasonable assumption of one of the quantities used, leading to an answer of a certain order of magnitude.

Remember the following points:
- State definitions accurately.
- Always write down the formula you are using in a calculation.
- Show all substitutions and working out.
- Check you have included the correct units.
- Use the correct number of significant figures.
- Use a ruler and pencil to draw simple diagrams accurately and neatly.
- Label diagrams fully.
- Know how to use your calculator — you will need it.

Revision tips

- Be familiar with the specification.
- Organise your notes and make sure they are complete.
- Learn all the equations indicated in the specification as those you must recall.
- Be familiar with the equations that are provided on the *Data and Formulae Sheet*.
- Practise rearranging and using equations to find different quantities.
- Learn definitions and laws thoroughly and accurately.
- Be able to describe all the experiments referred to in the specification with the aid of a labelled diagram.

Mathematics useful in Unit AS 2

- Standard form is a convenient way of writing down very large or very small numbers. The number is written as a number between 1 and 10 multiplied by 10 to the appropriate power — e.g. $345\,000 = 3.45 \times 10^5$. Be familiar with the prefixes that represent decimal multiples and submultiples of units:

Prefix	Symbol	Multiplying factor
kilo	k	10^3
mega	M	10^6
giga	G	10^9

Prefix	Symbol	Multiplying factor
milli	m	10^{-3}
micro	μ	10^{-6}
nano	n	10^{-9}

- Round numbers to the appropriate number of significant figures:
  - All non-zero digits are significant.
  - All zeros between significant digits are significant.
  - All zeros to the right of the decimal place are significant — for example, 0.0500 (3 s.f.), 9.00 (3 s.f.).
  - All zeros following significant digits in a whole number are not significant — for example 74 500 (3 s.f.).

- Degrees and radians:
  - The angle between two lines can be measured in degrees or radians. In a full circle (360°) there are 2π radians. Angles can be converted between degrees and radians as follows:

$$\text{angle in radians} = \frac{2\pi}{360} \times \text{angle in degrees}$$

- Sine and cosine graphs in terms of degrees and radians:

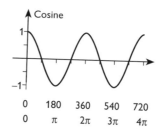

There will be a *Data and Formulae Sheet* inside the Unit AS 2 examination paper. The constants given are as follows:

- speed of light in a vacuum $c = 3.00 \times 10^8\,\text{m s}^{-1}$
- elementary charge $e = 1.60 \times 10^{-19}\,\text{C}$
- the Planck constant $h = 6.63 \times 10^{-34}\,\text{J s}$
- mass of electron $m_e = 9.11 \times 10^{-31}\,\text{kg}$
- mass of proton $m_p = 1.67 \times 10^{-27}\,\text{kg}$
- acceleration of free fall on the Earth's surface $g = 9.81\,\text{m s}^{-2}$
- electron volt $1\,\text{eV} = 1.60 \times 10^{-19}\,\text{J}$

The formulae given that are relevant to Unit AS 2 are as follows:

- sound 3 level/dB $10\log_{10}\dfrac{I}{I_0}$
- two-source interference $\lambda = \dfrac{ay}{d}$
- lens formula $\dfrac{1}{u}+\dfrac{1}{v}=\dfrac{1}{f}$
- magnification $m = \dfrac{v}{u}$
- de Broglie formula $\lambda = \dfrac{h}{p}$

About this section

This section consists of two self-assessment tests. Try the questions without looking at the answers, allowing 1 hour for each test. Then check your responses against the answers and the examiner comments to find out how you might improve your performance.

For question parts worth multiple marks, ticks (✓) are included in the answers to indicate where the examiner has awarded marks.

Examiner comments

Examiner comments on some questions are preceded by the icon 🄴. They offer tips on what you need to do to gain full marks. Some answers are followed by examiner comments, indicated by the icon 🄴, which highlight where credit is due or could be missed.

Unit 2: Waves, Photons and Medical Physics

Question 5

(a) **What are the basic principles of the production of X-rays?** (4 marks)

(b) **In what way does a CT image differ from a conventional X-ray image?** (4 marks)

(c) **State why MRI scanning is safer than CT scanning.** (2 marks)

(d) **State two problems associated with the use of MRI scanning.** (2 marks)

Total: 12 marks

Answer to Question 5

(a) Electrons from a hot cathode are accelerated ✓ in an evacuated chamber by a high voltage to a target metal anode ✓.
Two phenomena result in the production of X-rays: the sudden deceleration of the electrons ✓; and higher-level electrons dropping energy levels to fill vacancies left by electrons in inner shells being knocked out by the incident electrons ✓.

ⓔ This is a complicated concept. It is helpful when you are revising to learn to draw and label the X-ray tube and practise explaining the function of each part.

(b) The CT image is produced by a rotating X-ray tube and range of detectors that take multiple images of slices of part of the body ✓, which are combined by a computer to give a three-dimensional image ✓.
The conventional X-ray is produced by a stationary X-ray tube and a detector ✓, which produce a two-dimensional image of part of the body. ✓

ⓔ Comparison of conventional two-dimensional and three-dimensional X-rays is a common question. A comparison of the safety of the two types should prompt a discussion of the increased amount of dangerous ionising X-rays received in CT scans.

(c) CT scans use X-rays, which are ionising radiation and can damage living cells. ✓
MRI scans do not use ionising radiation but instead use radio waves and magnetic fields, which are not believed to cause significant damage to human tissue. ✓

ⓔ In questions that prompt a comparison, present details of both — do not assume that anything will be implied. Medical imaging questions often require a comparison to be made between imaging techniques.

(d) It is more costly than other imaging techniques as a strong magnetic field is required. This can be provided by superconducting magnets (shielding is also required for this magnetic field). ✓
Patients with metal implants are unable to be scanned with MRI due to the strong magnetic field. ✓

(b) Complete the figure below to show the type of lens used to correct myopia and show where the rays are now focused. Explain your completed diagram.   (2 marks)

(c) What component of the eye provides the greatest refracting power in the eye?   (1 mark)

(d) What is meant by the term accommodation in the context of eyesight?   (1 mark)

(e) Describe the mechanism of this process and explain why a malfunction of this process may lead to myopia and hypermetropia.   (3 marks)

Total: 9 marks

Answer to Question 3

(a)

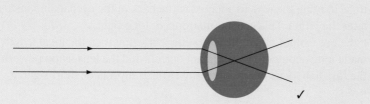

Myopia is short sight, when the eye cannot focus light from distant objects on the retina. The eye lens is too powerful and rays meet before retina. ✓

e Refraction takes place as the light enters the eye at the air–cornea boundary and at the lens. If possible show this on the diagram but as long as you show the overall converging of the rays that will suffice. Clearly show the rays crossing before the retina.

(b)

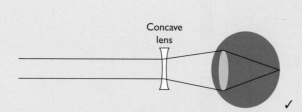

A concave (diverging) lens is used to spread the rays further apart, giving the eye lens more work to do and so bringing the rays to a point on the retina. ✓

ⓔ Make sure you learn that concave lenses help myopia and convex lenses help hypermetropia.

(c) The boundary between air and the cornea has the greatest difference in optical density and therefore the greatest refractive index. Hence the greatest amount of refraction takes place here in the eye, but this is a fixed amount of refraction. ✓

ⓔ It is not the eye lens. Always give as full an answer as possible.

(d) The ability of the eye to produce clear images of objects over a wide range of distances from the eye. ✓

ⓔ It is not just the ability to produce images. You must refer to clear or sharp images.

(e) The ciliary muscles in the eye stretch or squash the shape of the lens to make it thicker, with a short focal length, to view near objects clearly, and thinner, with a longer focal length, to view far objects clearly. ✓
If the eye cannot make the lens sufficiently thick then the focal length cannot be made short enough to view close objects as the lens is not as powerful as it needs to be to bring the rays to a focus on the back of the eyeball and instead they cross after the retina. This is hypermetropia or long sight. ✓
If the eye cannot make the lens sufficiently thin then the focal length cannot be made long enough to view distant objects and the lens is too powerful and brings the rays to a focus before the retina. This is myopia or short sight. ✓

ⓔ Don't confuse myopia and hypermetropia.

Question 4

Sea waves enter a harbour and are diffracted as they pass through the harbour gate.

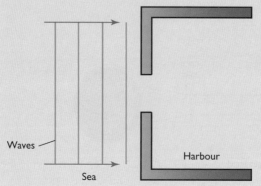

(a) Explain the meaning of diffraction.   (2 marks)

(b) Describe the conditions necessary for observable diffraction.   (1 mark)

(c) Add three more wavefronts to this diagram to illustrate the waves entering the harbour.   (2 marks)

(d) At another part of the coast the sea waves change direction and their wavelength decreases. Why does this happen?   (2 marks)

                                      Total: 7 marks

Answer to Question 4

(a) Diffraction is the spreading of waves ✓ when they pass through an opening or round an obstacle ✓.

🌐 Learn this definition of diffraction.

(b) For observable diffraction the width of the gap through which the wave is passing must be of approximately the same size as the wavelength of the wave. ✓

(c)

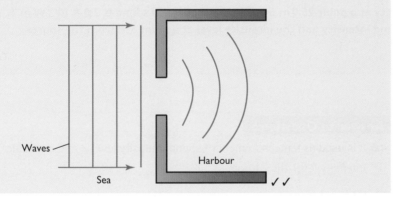

🌐 A mark will be awarded for the three wavefronts that have been requested if they are the correct shape. The second mark is for accuracy of your diagram, which must show that there is no change in wavelength, and you must indicate this by maintaining a constant distance between the wavefronts.

(d) Change of direction and decreasing wavelength mean that the wave has slowed down ✓ and has been refracted, so it must have entered shallower water at an angle ✓.

Question 5

(a) Explain fully what is meant by the decibel scale. (3 marks)

(b) (i) For a point source of sound, the intensity of sound at a distance from that point obeys an inverse square relationship. Hence the intensity of sound decreases with the square of the distance between the source and the point. Sketch a graph to show this relationship on the axes below. (2 marks)

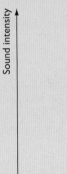

Sound intensity

Distance from
source of sound

(ii) The sound intensity at a point 25.0 m from a small source of sound is $2.0 \times 10^{-5} \, \text{W m}^{-2}$. Calculate the sound intensity and the intensity level at a point 5 m from the source. (3 marks)

Total: 8 marks

Answer to Question 5

(a) It is used as a measurement of sound intensity level ✓ and is a scale defined by:

intensity level/dB $= 10 \log_{10} \dfrac{I}{I_0}$ ✓

where I is the sound intensity and I_0 is the smallest audible sound intensity taken to be $1.0 \times 10^{-12} \, \text{W m}^{-2}$. ✓

ⓔ Any part of this answer will be awarded a mark but all three parts are necessary for full marks.

(b) (i)

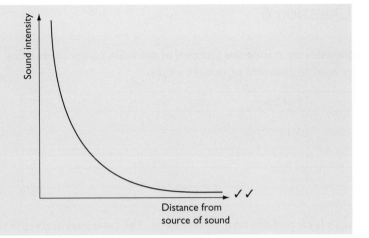

e This is the shape of a typical inverse square relationship. 1 mark is awarded for a curve and another mark for the curve being correct.

(ii) $\text{intensity} \propto \dfrac{1}{\text{distance}^2}$

e We know that the intensity obeys an inverse square relationship with the distance so we can write down a general proportionality equation for this. Removing the proportionality sign and replacing it with an equals sign gives us a constant k. By using the value of sound intensity at a distance of 25 m as $2.0 \times 10^{-5}\,\text{W}\,\text{m}^{-2}$, the value of the constant k can be found.

$$\text{intensity} = \frac{k}{\text{distance}^2}$$

$$2.0 \times 10^{-5} = \frac{k}{25^2} \checkmark$$

$$k = 2.0 \times 10^{-5} \times 25^2 = 0.0125$$

The intensity can be found for any distance using:

$$I = \frac{0.0125}{x^2}$$

For a distance of 5 m the intensity is given by:

$$I = \frac{0.0125}{5^2} = 5 \times 10^{-4}\,\text{W}\,\text{m}^{-2} \checkmark$$

e This is not the end of the question as the intensity level is also required.

$$\text{intensity level/dB} = 10\log_{10}\frac{I}{I_0}$$

$$\text{intensity level} = 10\log_{10}\frac{5 \times 10^{-4}}{10^{-12}} \checkmark$$

$$\text{intensity level} = 87\,\text{dB}$$

e Note the unit of intensity ($\text{W}\,\text{m}^{-2}$) and intensity level (dB).

Question 6

(a) Complete the following table by stating the function of the main components of an ultrasound transducer used to generate ultrasonic waves. *(5 marks)*

Component	Function
Piezoelectric crystal	
Acoustic insulator	
Backing material	
Thin plastic membrane	
Co-axial cable	

(b) (i) An ultrasonic A-scan is carried out on a human bone. The time base setting of the cathode ray oscilloscope is set at $10\,\mu s\,cm^{-1}$. A pulse of ultrasound is received as the ultrasound signal enters the bone and this produces a peak displayed on the screen. A second pulse is received as the signal leaves the bone and produces another peak on the screen. The distance between the peaks is 1.6 cm. Ultrasound travels at a speed of $4.0 \times 10^3\,m\,s^{-1}$ in bone. Calculate the thickness of the bone. *(3 marks)*

(ii) The frequency of the ultrasound used in this A-scan is 1.5 MHz. Calculate the wavelength of the ultrasound. *(2 marks)*

Total: 10 marks

Answer to Question 6

(a)

Component	Function
Piezoelectric crystal	Vibrates at source frequency when a.c. voltage applied to it ✓
Acoustic insulator	Prevents any external sounds affecting crystal ✓
Backing material	Damps crystal oscillations when applied voltage is stopped ✓
Thin plastic membrane	Transmits ultrasound from transducer ✓
Co-axial cable	Provides a.c. voltage to crystal ✓

ⓔ Distinguish between the acoustic insulator, which stops sound other than the ultrasonic signal from reaching the crystal, and the backing material, which damps the crystal vibrations to separate the ultrasonic signals.

CCEA AS Physics

(b) (i) The time between the peaks is the distance between the peaks multiplied by the time base setting:

time $= 1.6 \times 10 \times 10^{-6} = 1.6 \times 10^{-5}\,\text{s}$ ✓

speed $= 4.0 \times 10^{3}\,\text{m s}^{-1}$

distance $= $ speed $\times$ time $= 4.0 \times 10^{3} \times 1.6 \times 10^{-5} = 0.064\,\text{m}$ ✓

thickness of bone $= \tfrac{1}{2} \times 0.064\,\text{m} = 0.032\,\text{m}$ ✓

e There is always a factor of two to be taken into consideration in an echo question. 0.064 m is not the thickness of the bone. This is the distance from one side of the bone to the other and back again, as this is the extra distance the ultrasound has travelled to form the second peak.

(ii) $v = f\lambda$

$\lambda = \dfrac{v}{f} = \dfrac{4.0 \times 10^{3}}{1.5 \times 10^{6}}$ ✓

$\lambda = 0.0026\,\text{m}$ ✓

Question 7

Some of the energy levels for the mercury atom are −3.70 eV, −5.51 eV and −10.40 eV (ground state).

(a) Calculate the energy required to excite an electron in the ground state to each of the other two levels. (2 marks)

(b) Describe what would happen if cold mercury is bombarded with:
(i) electrons of kinetic energy 2.5 eV (2 marks)
(ii) electrons of kinetic energy 6 eV (2 marks)
(iii) light of wavelength 254 nm (3 marks)
(iv) light of wavelength 200 nm (3 marks)

Total: 12 marks

Total for this self-assessment test = 70 marks

Answer to Question 7

(a) It would take $(-5.51\,\text{eV}) - (-10.40\,\text{eV}) = 4.89\,\text{eV}$ to raise an electron from the ground state to the −5.51 eV energy level. ✓
It would take $(-3.70\,\text{eV}) - (-10.40\,\text{eV}) = 6.70\,\text{eV}$ to raise an electron from the ground state to the −3.70 eV energy level. ✓
(b) (i) The atom needs at least 4.89 eV to raise it from the ground state to a higher level. 2.5 eV is insufficient. ✓
The bombarding electrons collide with the electrons in the atoms, not giving up any of their energy, and move on. ✓

(ii) The atom needs at least 4.89 eV to raise it from the ground state to a higher level. 6 eV is more than sufficient. ✓
So the electron gives up 4.89 eV of energy and comes away with 1.11 eV. The excited atom will return to the ground state with the emission of a photon of energy 4.89 eV. ✓

(iii) Photons can only give up all or nothing of their energy.
A wavelength of 254 nm can be converted to energy using:

$$E = \frac{hc}{\lambda} = \frac{6.63 \times 10^{-34} \times 3 \times 10^8}{254 \times 10^{-9}}$$

$$= 7.83 \times 10^{-19}\,\text{J} = \frac{7.83 \times 10^{-19}}{1.6 \times 10^{-19}}\,\text{eV} = 4.89\,\text{eV}\ ✓$$

This is exactly the correct amount of energy so it is the exact wavelength needed and the photon is absorbed. ✓ The atom returns to the ground state with the release of a photon of 4.89 eV. ✓

(iv)

$$E = \frac{hc}{\lambda} = \frac{6.63 \times 10^{-34} \times 3 \times 10^8}{200 \times 10^{-9}}$$

$$= 9.945 \times 10^{-19}\,\text{J} = \frac{9.945 \times 10^{-19}}{1.6 \times 10^{-19}} = 6.22\,\text{eV}\ ✓$$

This is too much energy to promote an electron from the ground state to the first energy level and too little to promote an electron from the ground state to the second level. ✓
A photon must transfer all or none of its energy so the photons are not absorbed. ✓

ⓔ It is important to understand the difference between a photon providing the energy to electrons and the electrons receiving it through collisions with other electrons (or by heat). Photons are packets of electromagnetic energy and only exist as whole entities — they cannot give up part of their energy, just all or nothing.

Self-assessment test 2

Question 1

(a) List three properties common to all electromagnetic waves. (3 marks)

(b) Explain why the electromagnetic spectrum is described as continuous. (1 mark)

(c) State four electromagnetic waves that have shorter wavelengths than indigo light. (2 marks)

(d) Calculate the wavelength of radio waves of frequency 900 kHz. (3 marks)

(e) Describe and explain how light can be plane-polarised and suggest a method to check polarisation has taken place. (4 marks)

Quality of written communication (2 marks)

Total: 15 marks

Answer to Question 1

(a) Electromagnetic waves:
- travel at the speed of light ✓
- are transverse waves ✓
- can travel through a vacuum ✓
- (*or* consist of oscillating electric and magnetic fields ✓)

ⓔ A common mistake is suggesting that electromagnetic waves are longitudinal waves. This is totally incorrect.

(b) There is a wave at every wavelength — there are no gaps. The waves overlap as there are no clear boundaries between the waves. ✓

(c) Violet light, ultraviolet radiation, X-rays, gamma rays ✓ ✓ (½ each)

ⓔ Know the electromagnetic spectrum in order of increasing wavelength and be able to suggest typical wavelengths for each wave in the spectrum.

(d) $f = 900\,\text{kHz} = 9 \times 10^5\,\text{Hz}$ ✓

$c = 3.0 \times 10^8\,\text{m s}^{-1}$

$v = f\lambda$ ✓

For electromagnetic waves $c = f\lambda$

$\lambda = \dfrac{c}{f} = \dfrac{3 \times 10^8}{9 \times 10^5} = 333\,\text{m}$ ✓

ⓔ Check the units to convert from multiples such as kilohertz. Write the formula in the correct form (changing the subject if necessary) before making substitutions.

> **(e)** In unpolarised light the oscillations occur in all planes. ✓
> When unpolarised light is passed through a polarising filter such as Polaroid the oscillations in all planes but one will be absorbed and the oscillations will be in one plane only. ✓
> A second polariser is used as an analyser to confirm that the light is plane polarised. ✓
> The analyser is rotated through 360° and the light intensity will be extinguished twice in the rotation. ✓

ⓔ Describe experiments or practical procedures carefully, with all the details. For example, don't just say an analyser is used to check polarisation. Describe in detail *how* it is used.

ⓔ Quality of written communication will be assessed at some point during an examination. It will say at the beginning of the examination paper which question will include this element. For this question, use complete sentences with capital letters, full stops, accurate spelling and legible text.

Question 2

> The figure shows a ray of light incident at an angle of 75° on a glass block. The ray enters the glass and meets the side BC. The critical angle for glass is 41°.
>
>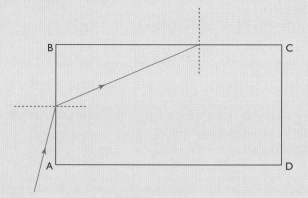
>
> **(a) (i)** Explain what is meant by the term refraction of light.   (1 mark)
>   **(ii)** Explain what is meant by the term absolute refractive index.   (2 marks)
>
> **(b)** Calculate the refractive index of the glass.   (2 marks)
>
> **(c)** Show that the ray will not emerge from the block at side BC.   (3 marks)
>
> **Total: 8 marks**

Answer to Question 2

(a) (i) Refraction is the bending of light as it travels between materials of different densities and changes speed. ✓

(ii) Refractive index is a measure of the change of direction of light as it travels from air (or a vacuum) to another transparent material. ✓

It is the ratio of the speed of the light in air (or a vacuum) to the speed of the light in the second medium. ✓

(Or it is the ratio of the sine of the angle of incidence to the sine of the angle of refraction across the boundary between air (or a vacuum) and the second medium. ✓)

🅔 To be awarded 2 marks a full description is required.

(b)

$$n = \frac{1}{\sin C}$$

$$n = \frac{1}{\sin 41} \; ✓$$

$$n = 1.52 \; ✓$$

(c) Find the angle of refraction:

$$n = \frac{\sin i}{\sin r}$$

$$\sin r = \frac{\sin 75}{1.52}$$

$$r = 39.5 \; ✓$$

Use this to find the angle of incidence at BC by considering the geometry of the block.

angle of incidence at BC = 50.5° ✓

This is greater than the critical angle so total internal reflection will take place and the light will not emerge from the block. ✓

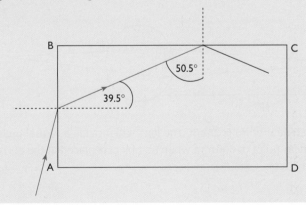

Questions & Answers

It is not necessary to measure angles accurately with a protractor in this question. However, as you calculate the angles using Snell's relationship and the geometry of a triangle, mark the angles on the diagram and make sure they are a good approximation to the actual value. Explain your reasoning as you work through the question and show all working out.

Question 3

(a) State the meaning of hypermetropia. (1 mark)

(b) State the type of lens used to correct hypermetropia. (1 mark)

(c) (i) A person with hypermetropia has a near point of 600 mm. Complete the figure below to show how light from the person's near point would be focused by their unaided eye. (2 marks)

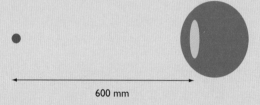

600 mm

(ii) Calculate the power of the lens required to correct the person's near point to a normal distance. (3 marks)

Total: 7 marks

Answer to Question 3

(a) Distant objects can be focused but close objects cannot. ✓
(b) Convex — converging lens. ✓
(c) (i)

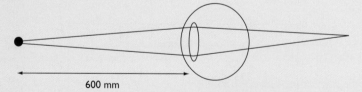

600 mm

(ii) A normal near point is 25 cm. The lens will provide a virtual image at the actual near point of 600 mm when an object is placed at the corrected near point of 25 cm.

$$\frac{1}{f} = \frac{1}{u} + \frac{1}{v}$$

$$\frac{1}{f} = \frac{1}{250} + \frac{1}{-600}$$

$$f = 429 \, \text{mm} = 0.43 \, \text{m} ✓$$

$$P = \frac{1}{f} = \frac{1}{0.43} = 2.3 \, \text{D} ✓$$

CCEA AS Physics

❸ Remember to convert all quantities to the same units. Note that the question asks for the power and if dioptres is the unit then metres must be used for the focal length.

Question 4

(a) What is meant by the term sound intensity?   (1 mark)

(b) A girl can hear a minimum sound intensity of $1.0 \times 10^{-12}\,W\,m^{-2}$ at a frequency of 2000 Hz. The area of the entrance to her ear is $50\,mm^2$. Calculate the minimum sound power that she can just detect at 1000 Hz.   (3 marks)

(c) Discomfort or pain can be induced by sounds of intensity level 120 dB. Calculate the sound intensity corresponding to this intensity level.   (3 marks)

                                        Total: 7 marks

Answer to Question 4

(a) The power passing through unit area at right angles to the direction of the sound wave. ✓

(b) intensity = $\dfrac{\text{power}}{\text{area}}$

So,

power = intensity × area ✓

area = $50\,mm^2 = 50 \times 10^{-6}\,m^2$ ✓

power = $1.0 \times 10^{-12} \times 50 \times 10^{-6} = 5 \times 10^{-17}\,W$ ✓

❸ Don't panic if you think you don't know a formula. The formula for intensity can be obtained directly from the units of intensity, $W\,m^{-2}$.

(c) intensity level = $10 \log_{10} \dfrac{I}{I_0}$

$= 10 \log_{10} \dfrac{I}{10^{-12}} = 120\,dB$ ✓

$\log_{10} \dfrac{I}{10^{-12}} = \dfrac{120}{10}$

$\log_{10} \dfrac{I}{10^{-12}} = 12$ ✓

$\dfrac{I}{10^{-12}} = 10^{12}$

$I = 10^{12} \times 10^{-12} = 10^0 = 1\,W\,m^{-2}$ ✓

❸ Show all steps of your calculation. Note that $I_0 = 1.0 \times 10^{-12}\,W\,m^{-2}$, which is taken as the threshold intensity at frequencies of about 2 kHz. This is important because it is used in intensity level calculations.

Question 5

(a) What are the basic principles of the production of X-rays? (4 marks)

(b) In what way does a **CT** image differ from a conventional **X-ray** image? (4 marks)

(c) State why **MRI** scanning is safer than **CT** scanning. (2 marks)

(d) State two problems associated with the use of **MRI** scanning. (2 marks)

Total: 12 marks

Answer to Question 5

(a) Electrons from a hot cathode are accelerated ✓ in an evacuated chamber by a high voltage to a target metal anode ✓.
Two phenomena result in the production of X-rays: the sudden deceleration of the electrons ✓; and higher-level electrons dropping energy levels to fill vacancies left by electrons in inner shells being knocked out by the incident electrons ✓.

ⓔ This is a complicated concept. It is helpful when you are revising to learn to draw and label the X-ray tube and practise explaining the function of each part.

(b) The CT image is produced by a rotating X-ray tube and range of detectors that take multiple images of slices of part of the body ✓, which are combined by a computer to give a three-dimensional image ✓.
The conventional X-ray is produced by a stationary X-ray tube and a detector ✓, which produce a two-dimensional image of part of the body. ✓

ⓔ Comparison of conventional two-dimensional and three-dimensional X-rays is a common question. A comparison of the safety of the two types should prompt a discussion of the increased amount of dangerous ionising X-rays received in CT scans.

(c) CT scans use X-rays, which are ionising radiation and can damage living cells. ✓
MRI scans do not use ionising radiation but instead use radio waves and magnetic fields, which are not believed to cause significant damage to human tissue. ✓

ⓔ In questions that prompt a comparison, present details of both — do not assume that anything will be implied. Medical imaging questions often require a comparison to be made between imaging techniques.

(d) It is more costly than other imaging techniques as a strong magnetic field is required. This can be provided by superconducting magnets (shielding is also required for this magnetic field). ✓
Patients with metal implants are unable to be scanned with MRI due to the strong magnetic field. ✓

CCEA AS Physics

Question 6

The figure shows the ground state and some of the excited energy levels in the hydrogen atom.

```
0        ────────────────
-0.54    ──────────────── E
-0.85    ──────────────── D

-1.51    ──────────────── C
E/eV

-3.40    ──────────────── B


-13.59   ──────────────── A
```

(a) To excite an electron from the ground state to a higher level requires energy. Where might this energy come from?   (3 marks)

(b) How much energy is required to move an electron from the ground state out of the atom?   (1 mark)

(c) Why do the energy levels have negative values?   (2 marks)

(d) What type of spectrum results from transitions of the electron between energy levels in the atom?   (1 marks)

(e) Calculate the wavelength of radiation emitted by a transition between level D and level B and state the region of the electromagnetic spectrum corresponding to this wavelength.   (3 marks)

(f) Laser action depends on the stimulated emission of radiation. Explain in terms of energy levels, what is meant by the term 'stimulated emission'.   (3 marks)

Total: 13 marks

Answer to Question 6

(a) The electrons can gain energy by heating, by collision or by absorption of photon energy. ✓✓✓

(b) 13.59 eV ✓

(c) The energy of an electron just outside the atom is taken as zero ✓ and if energy is added to get the electron out of the atom then the energy levels must be negative. ✓

(d) Line spectrum ✓

(e)

change in energy $= (-0.85\,\text{eV}) - (-3.40\,\text{eV}) = 2.55\,\text{eV} = 2.55 \times 1.6 \times 10^{-19}\,\text{J}$ ✓ $= \dfrac{hc}{\lambda}$

$\lambda = \dfrac{6.63 \times 10^{-34} \times 3 \times 10^{8}}{2.55 \times 1.6 \times 10^{-19}} = 4.88 \times 10^{-7}\,\text{m}$

$= 488\,\text{nm}$ ✓ $=$ visible light ✓

ⓔ Remember it is E_2 (the higher energy level) $- E_1$ (the lower energy level) = the energy of the emitted photon. Keep the minus signs in the correct positions to obtain the correct result. Always convert from eV to joules to use formulae.

> **(f)** The atom is initially in an excited state, with electrons in higher energy levels. ✓ An incident photon of exactly the right energy causes an electron to fall to a lower level ✓ with the emission of a photon of the same energy and in phase with the incident one ✓.

ⓔ This part of the question is only awarded 3 marks, so only a brief description is required, focused on the energy levels. Another question with a higher mark allocation might require a more detailed description.

Question 7

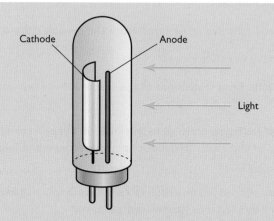

A photocell is a device that can be used to measure the intensity of light incident upon it. When photons of light hit the cathode, some electrons may be released. This emission depends on the energy of the photon and the work function of the cathode.

(a) Explain the meaning of the term work function and state the conditions under which photoelectric emission will take place.
(2 marks)

(b) Caesium has a work function of 1.9 eV. Calculate the lowest frequency of radiation that will produce photoelectric emission from a cathode containing caesium.
(3 marks)

(c) If red light shines on the cathode will this cause photoelectric emission?
(3 marks)

Total: 8 marks

Total for this self-assessment test = 70 marks

Answer to Question 7

(a) The work function of a material is the minimum energy required to remove an electron from the surface of that material. ✓

Photoelectric emission will take place if electromagnetic radiation incident on the surface of the material provides photons of energy greater than or equal to the work function. ✓

ⓔ This question is about the work function of a metal. Do not answer in terms of threshold frequency.

(b) Convert eV to joules.

$1.9 \, eV = 1.9 \times 1.6 \times 10^{-19} = 3.04 \times 10^{-19} \, J$ ✓

The work function is given by Planck's constant times the threshold frequency:

$\phi = h f_0$

So

$f_0 = \dfrac{\phi}{h} = \dfrac{3.04 \times 10^{-19}}{6.63 \times 10^{-34}}$ ✓

$f_0 = 4.6 \times 10^{14} \, Hz$ ✓

(c) Red light has a wavelength of about 700 nm. ✓

So the frequency of red light is:

$f = \dfrac{c}{\lambda} = \dfrac{3.0 \times 10^8}{700 \times 10^{-9}} = 4.3 \times 10^{14} \, Hz$ ✓

This is not a high enough frequency to cause photoemission. ✓

ⓔ Be able to suggest typical wavelengths for all electromagnetic waves including all the visible light waves.

Knowledge check answers

1 In transverse waves the vibrations are at right angles to the direction of travel of the wave. Examples include any electromagnetic wave and water waves. In longitudinal waves the vibrations are parallel to the direction of travel of the wave. Examples include sound, ultrasound and seismic p-waves.

2 Frequency can be determined from a displacement–time graph. The time between successive crests will give the periodic time. This can be used to find the frequency, as periodic time is $\frac{1}{\text{frequency}}$ (or $f = \frac{1}{T}$).

Frequency cannot be determined from a displacement-distance graph. Wavelength is taken as the distance between successive crests and if the speed of the wave is known only then can the frequency be determined by using the wave equation.

3 $T = 1\,\text{ms} = 0.001\,\text{s}$ (change milliseconds to seconds)
frequency $= \frac{1}{\text{period}} = \frac{1}{0.001} = 1000\,\text{Hz}$

4 The wave equation is $v = f\lambda$.
$v = 68000 \times 0.005 = 340\,\text{ms}^{-1}$

5 phase difference $= \frac{3.5}{14} \times 360° = 90°$ or $\frac{\pi}{2}$ radians

6 The light would vary gradually from a maximum to zero intensity twice in the rotation.

7 Radio waves travel at the speed of light, can travel through a vacuum and are transverse waves. Radio waves are not considered as dangerous as they have a long wavelength, low frequency or low energy, and are not ionising radiation.

8 The speed will increase, the wavelength will increase and the frequency will stay the same.

9 Proportionality — the graph is a straight line through the origin and as you double the dependent variable, the independent variable also doubles.

10 $n = 1.5 = $ speed of light/speed of light in glass $= \frac{3 \times 10^8}{c_g}$
$c_g = 2.0 \times 10^8\,\text{ms}^{-1}$

11 Light is faster than electricity; there is less interference of signal; the raw materials are cheaper; more information can be carried.

12 A point on the principal axis through which rays of light travelling parallel to the principal axis converge after refraction by the lens.

13 Erect (upright), enlarged (magnified), virtual (on the same side of the lens as the object).

14 $\frac{1}{f} = \frac{1}{4} + \frac{1}{-8} = \frac{1}{8}$ $f = 8\,\text{cm}$

15 The diverging lens is thin in the middle, the convex lenses are thick in the middle and the fatter convex lens has the shortest focal length.

16 $\frac{1}{v} = -\frac{1}{u} + \frac{1}{f}$

if $\frac{1}{v} = 0$ then $\frac{1}{u} = \frac{1}{f}$

if $\frac{1}{u} = 0$ then $\frac{1}{v} = \frac{1}{f}$

17 A powerful lens has a short focal length and is a thick lens.

18 The greatest amount of refraction in the eye takes place at the air–cornea boundary as there is a large difference in refractive index between the air and the cornea. Under water there is little difference between the water and the cornea and less refraction takes place when light enters the eye. The eye lens is not sufficiently powerful to focus the light onto the retina itself.

19 Normal near point to normal far point: 25 cm to infinity.

20 Waves that meet must be coherent, have the same amplitude and be out of phase by 180°.

21 Standing — the wave does not transfer energy from one place to another.
Mechanical — the wave is generated by disturbances in a material medium.
Longitudinal — vibrations parallel to direction of travel; there are areas of compression and rarefaction.

22 The fringe separation increases.

23 A sound level meter could be adjusted to respond more to sounds between 2 kHz and 4 kHz — a frequency-weighted scale modelled on the ear's response.

24 If surgery is used the patient has a longer recovery time, and may have complications due to the anaesthetic and the risk of infection.

25 The image must be transmitted with a coherent bundle because the fibres will maintain a constant position and so the image will be an accurate copy of the object. This is not required for illumination purposes as randomly orientated fibres will still enable the light to get to the target.

26 The acoustic insulator prevents sounds signals and vibrations from sources other than the imaging signal affecting the piezoelectric crystal.

27 The tungsten target absorbs the electrons and releases some of the energy in the form of X-rays by the rapid deceleration of electrons after passing the nucleus or by the tightly bound inner electrons being knocked out of the atom by the incident electrons and higher-state electrons dropping down to fill the vacancy.

28 Quantised refers to discrete units or whole packets of energy.